T0225274

SERIES ON

Water: Emerging Issues and Innovative Responses

WATER, SUSTAINABLE DEVELOPMENT AND THE NEXUS

Response to Climate Change

Editors

Velma I. Grover

Adjunct Professor
Faculty of Environmental Studies
York University
Toronto, ON, Canada

Amani Alfarra

Water Resource Officer
CBL Division – FAO of the U.N.
Rome, Italy

CRC Press

Taylor & Francis Group
Boca Raton London New York

CRC Press is an imprint of the
Taylor & Francis Group, an **informa** business

A SCIENCE PUBLISHERS BOOK

Cover illustration: Wordclouds.com was used to generate the image.

CRC Press
Taylor & Francis Group
6000 Broken Sound Parkway NW, Suite 300
Boca Raton, FL 33487-2742

First issued in paperback 2021

© 2020 by Taylor & Francis Group, LLC
CRC Press is an imprint of Taylor & Francis Group, an Informa business

No claim to original U.S. Government works

Version Date: 20190516

ISBN-13: 978-0-367-77667-1 (pbk)
ISBN-13: 978-1-4987-8651-5 (hbk)

This book contains information obtained from authentic and highly regarded sources. Reasonable efforts have been made to publish reliable data and information, but the author and publisher cannot assume responsibility for the validity of all materials or the consequences of their use. The authors and publishers have attempted to trace the copyright holders of all material reproduced in this publication and apologize to copyright holders if permission to publish in this form has not been obtained. If any copyright material has not been acknowledged please write and let us know so we may rectify in any future reprint.

Except as permitted under U.S. Copyright Law, no part of this book may be reprinted, reproduced, transmitted, or utilized in any form by any electronic, mechanical, or other means, now known or hereafter invented, including photocopying, microfilming, and recording, or in any information storage or retrieval system, without written permission from the publishers.

For permission to photocopy or use material electronically from this work, please access www.copyright.com (http://www.copyright.com/) or contact the Copyright Clearance Center, Inc. (CCC), 222 Rosewood Drive, Danvers, MA 01923, 978-750-8400. CCC is a not-for-profit organization that provides licenses and registration for a variety of users. For organizations that have been granted a photocopy license by the CCC, a separate system of payment has been arranged.

Trademark Notice: Product or corporate names may be trademarks or registered trademarks, and are used only for identification and explanation without intent to infringe.

Library of Congress Cataloging-in-Publication Data
Names: Grover, Velma I., editor. | Alfarra, Amani, editor.
Title: Water, sustainable development, and the nexus : response to climate change / editors, Velma I. Grover (adjunct professor, Faculty of Environmental Studies. York University, Toronto, ON, Canada), Amani Alfarra (water resource officer, CBL Division, FAO of the U.N., Rome, Italy).
Description: Boca Raton, FL : CRC Press, 2019. | Series: Series on water: emerging issues and innovative responses | "A science publishers book." | Includes bibliographical references and index.
Identifiers: LCCN 2019008175 | ISBN 9781498786515 (hardback)
Subjects: LCSH: Water-supply--Forecasting. | Climatic changes.
Classification: LCC TD390 .W3845 2019 | DDC 333.91--dc23
LC record available at https://lccn.loc.gov/2019008175

Visit the Taylor & Francis Web site at
http://www.taylorandfrancis.com

and the CRC Press Web site at
http://www.crcpress.com

Preface to the Series

Water is the lifeline for all life on the planet earth and is linked to the development of all societies and cultures. But this development of all societies and cultures comes with a tag for increased water demand: because of increasing population as well as because of increasing demand from competing sectors such as agriculture, industry, domestic consumption, and recreation. This has impacted both the water quality and water quantity. The planet is at a point where anthropogenic changes have started to interfere with natural processes, including water cycle, and ecosystem services. More innovative approaches will be required to deal with emerging problems and transforming challenges into opportunities. This involves finding innovative solutions of treating water pollution, better governance structure and other institutional framework, good legal structure, in addition to increased investment in water sector.

Water is at the core of sustainable development hence sustainable water management and governance and is vital for humanity and sustainable future of human society (including human health, environmental health, poverty reduction and economic growth). This water series, "Emerging Issues and Innovative Responses" is an attempt towards coming up with a comprehensive series which will look at various aspects of water science, management, governance, law, economics and its relation to (sustainable) development (including social and cultural attitudes), impact of climate change on water resources, including water, energy and food nexus.

Velma I. Grover

Preface

Water, food and energy are all central to sustainable development and are also inextricably linked to each other. Hence they require an integrated approach – the water-food-energy nexus approach – to deal with water, food and energy security issues. A lot of researchers are trying various nexus combinations that would include water, food, energy, trade, climate change etc – however the focus of this book is on "water-energy-food" nexus. Simply put, food production needs water (water extraction, treatment and distribution needs energy) as well as fertilizers (production of which needs energy), and energy production requires water. Since the three are so interlinked, addressing only one of these resources will complicate the problem for the other two sectors even more.

The nexus approach aims to identify tradeoffs and synergies of water, energy, and food systems, internalize social and environmental impacts, and guide the development of cross-sectoral policies. However, while the WEF nexus offers a promising conceptual approach, the use of WEF nexus methods to systematically evaluate water, energy, and food interlinkages or support development of socially and politically-relevant resource policies have been limited.

The introductory section of the book focuses on some concepts of water, energy and food security and nexus, economics of WEF nexus and demand management, followed by its application in various regions including North America, Europe, Africa and the Middle East in section two. Other sections focus on water as a human right, water security and the importance of education. This book is an attempt to look at the nexus approach and its applications in different continents.

Velma I. Grover
Amani Alfarra

Contents

Section 1
Introductory Section

Chapter-1

The Water-Energy-Food Nexus

Velma I. Grover[a],* and *Amani Alfarra*[b]

INTRODUCTION

"With global population growing over seven billion, accompanied with escalating economic crises, mismanagement of natural resources, climatic changes and uncertainties, and growing poverty and hunger, the world is living a critical period defined by global challenges. These challenges are linked with social, economic, and political risks and unrest that faces present and future generations" (Mohtar and Daher, 2012).[1] We face the challenge of effectively managing natural resources to better achieve sustainable outcomes, for example, those emphasized by the global community in the Sustainable Development Goals (SDGs, UN, 2015).[2] Just to put into perspective the latest numbers for food (only one aspect of the nexus), recent UN-multi-agency publication indicated that almost 800 million people are currently hungry. With the projected population of about 9 billion by 2050, global food production would need to increase by 50% (FAO/IFAD/UNICEF/WFP/WHO, 2017).[3] This would require so much more water and energy. "Different nexus combinations including water, food, energy, trade, climate, and population growth are being studied in an attempt to identify the types of interconnectedness present between those systems. The creation of those nexuses comes as a result of realizing the multi-dimensional and interdisciplinary nature of the foreseen challenges" (Mohtar and Daher, 2012).[4] In recent years, the Water-Energy-Food (WEF) nexus has become a lens to better understand complex

[a] 981 Main St West, #916, Hamilton, ON, L8S1A8, Canada.
[b] Water Resource Officer, CBL Division – Room B723 bis, FAO of the U.N., Viale delle Terme di Caracalla, 00153 Rome, Italy.
* Corresponding author: velmaigrover@yahoo.com

[1] Rabi H. Mohtar and Daher, B. 2012. Water, Energy, and Food: The Ultimate Nexus. Encyclopedia of Agriculture, Food and Biological Engineering, Second Edition. Taylor and Francis.
[2] United Nations (UN). 2015. Transforming our world: the 2030 Agenda for Sustainable Development. https://sustainabledevelopment.un.org/post2015/transformingourworld.
[3] http://www.fao.org/3/a-I7695e.pdf.
[4] Rabi H. Mohtar and Daher, B. 2012. Water, Energy, and Food: The Ultimate Nexus. Encyclopedia of Agriculture, Food and Biological Engineering, Second Edition. Taylor and Francis.

interactions among multiple resource systems. The focus of this book is on "water-energy-food" nexus. The nexus approach aims to identify tradeoffs and synergies of water, energy, and food systems, internalize social and environmental impacts, and guide the development of cross-sectoral policies. However, while the WEF nexus offers a promising conceptual approach, the use of WEF nexus methods to systematically evaluate water, energy, and food interlinkages or support development of socially and politically-relevant resource policies have been limited.

The introductory section of the book focuses on some concepts of water, energy and food security and nexus, economics of WEF nexus and demand management, followed by its application in various regions including North America, Europe, Africa and the Middle East in section two. Other sections focus on water as a human right, water security and the importance of education. By understanding the WEF nexus as a system-based perspective that explicitly recognizes water, energy, and food systems as both interconnected and interdependent (Bazilian et al., 2011; Wolfe et al., 2016; Foran, 2015).[5,6,7]

This chapter describes the interlinkages between water scarcity, food security and reviews different existing frameworks and approaches that deal with the interactions between different sectors followed by a road map for the book.

Water, Energy and Food Linkages and Nexus

One of the greatest challenges of the current century is rapidly increasing population coupled with increased demand in limited resources. It is projected that about 90% of population increase by 2030 will be in emerging markets followed by urbanization and industrialization in these economies accelerating demand for water, energy and food.[8] Water, energy and food are very intricately linked and the interdependence of these three resources is known as water-food-energy nexus.[9] The projected growth of population and increased demand of these resources in many countries makes one wonder where, would the business as usual scenario lead us. Most likely this will make it difficult to sustain economic and business growth to maintain adequate standard of living.[10] In very simple terms, food production (agriculture) needs water

[5]   Bazilian, M., Morgan Bazilian, Holger Rogner, Mark Howells, Sebastian Hermann, Douglas Arent, Dolf Gielen, Pasquale Steduto, Alexander Mueller, Paul Komor, Richard S.J. Tol, Kandeh K. Yumkella. 2011. Considering the energy, water and food nexus: towards an integrated modelling approach. *Energy Policy* 39: 7896–906.

[6]   Wolfe, M.L., Ting, K.C., Scott, N., Sharpley, A., Jones, J.W. and Verma, L. 2016. Engineering solutions for food-energy-water systems: it is more than engineering. *Journal of Environmental Studies and Science* 6: 172–82.

[7]   Foran, T. 2015. Node and regime: interdisciplinary analysis of water-energy-food nexus in the Mekong region. *Water Alternatives* 8: 655–74. http://www.water-alternatives.org/index.php/all-abs/270-a8-1-3/file.

[8]   Deflecting the scarcity trajectory: innovation at the water, energy and food nexus. Deloitte Review, issue 17. Will Sarni. Accessed from https://www2.deloitte.com/insights/us/en/deloitte-review/issue-17/water-energy-food-nexus.html.

[9]   UN Water. Water, food, and energy nexus. Accessed from http://www.unwater.org/topics/water-food-and-energy-nexus/en/.

[10] Deflecting the scarcity trajectory: innovation at the water, energy and food nexus. Deloitte Review, issue 17. Will Sarni. Accessed from https://www2.deloitte.com/insights/us/en/deloitte-review/issue-17/water-energy-food-nexus.html.

as well as fertilizers (production of which needs energy), while water extraction, treatment and distribution needs energy and energy production requires water.[11] Since the three are so interlinked, addressing only one of these resources will complicate the problem for the other two sectors even more. As Cramwinckel of World Business Council on Sustainable Development has rightly said, "Water, energy and food are intrinsically interrelated: A sustainable solution for one almost always has an impact on the others".[12] The economics of this nexus is becoming very clear in cases of Brazil, China and California in the US. For example, in California in the US, severe water scarcity has led to new rationing measures which impacts both food production and energy production (especially hydroelectric power) leading to economic impact on both private businesses as well as state economy. Reports have suggested that severe water scarcity/drought has led to an economic loss of US$2.2 billion in 2014 and somewhere around US$3 billion in 2015.[13] However, in every problem lies an opportunity and the chapters in the book explores how this nexus can be innovatively used to find solutions via partnership, technology, etc., to solve this complex issue.

Various authors and institutions have taken a slightly different approach towards nexus. The World Economic Forum has been among the first organizations to identify the water-energy-food nexus as a key development challenge, calling for a better understanding of the inter-linkages between water, energy and food at the 2008 Annual Meeting in Davos.[14] The World Economic Forum Water Initiative subsequently published a book, exploring the topic of water security in relation to energy and food systems as well as climate, economic growth and human security and the Water Resources Group launched a nexus initiative with water security as a practical entry point.[15] In line with this, the Bonn Nexus Conference in 2011 also emphasized the centrality of water resources.[16]

Some studies have also explored the impact of economic incentives in the water-food-energy nexus and have concluded that economic incentives can be a critical tool to manage the resources in the nexus. Economic growth leads to increased demand of water, energy and food but resource depletion and/or scarcity is barrier

---

[11] Rabi H. Mohtar and Daher, B. 2012. Water, Energy, and Food: The Ultimate Nexus. Encyclopedia of Agriculture, Food and Biological Engineering, Second Edition. Taylor and Francis.

[12] World Business Council for Sustainable Development. Joppe Cramwinckel's concluding remarks at the Founders Business Seminar during World Water Week in Stockholm. http://www.wbcsd.org/Pages/EDocument/EDocumentDetails.aspx?ID=13674& NoSearchContextKey=true AND Deflecting the scarcity trajectory: innovation at the water, energy and food nexus. Deloitte Review, issue 17. Will Sarni. Accessed from https://www2.deloitte.com/insights/us/en/deloitte-review/issue-17/water-energy-food-nexus.html.

[13] Jeff Daniels. 2014. California drought was bad. 2015 will be worse. CNBC, March 3, 2015, http://www.cnbc.com/id/102472848 and Deflecting the scarcity trajectory: innovation at the water, energy and food nexus. Deloitte Review, issue 17. Will Sarni. Accessed from https://www2.deloitte.com/insights/us/en/deloitte-review/issue-17/water-energy-food-nexus.html.

[14] WEF (ed.). 2011. Water Security: Water-Food-Energy-Climate Nexus. The World Economic Forum Water Initiative (Island Press, Washington D.C., USA).

[15] WEF. 2011. Global Risks 2011. 6th Edition (World Economic Forum (WEF), Cologne/Geneva).

[16] Hoff, H. 2011. Understanding the Nexus. Background Paper for the Bonn2011 Conference: The Water, Energy and Food Security Nexus (Stockholm Environment Institute (SEI), Stockholm, Sweden).

to economic growth.[17,18] As explored in some of the chapters in the book and some authors, energy and water prices are linked which are linked to food prices and food security.[19] Governance has also come up in literature as one of the challenges (or opportunities). Since, "governance challenges are at the heart of the nexus" (p.44)[20] the WEF nexus can be only be addressed by taking into account issues of institutional arrangements at the level where financial decisions are made.[21,22] Various authors have touched upon different aspects of governance as discussed by Galaitsi (2018): there is discussion around engaging stakeholders on equal footing, managing power imbalances; "optimal governance from a nexus perspective involves multi-level institutions and efforts at policy integration/coherence" (p.12); raising awareness of nexus among decision makers, etc.[23,24,25,26] The emphasis is on engaging stakeholders in the decision-making process at various levels for better understanding of nexus at all levels too.

Kurian (2018) has explored the use of water-food-energy nexus to enhance the societal relevance of global public goods research. The author has combined the concept of planetary boundaries as well in this approach. This would support "definition of the Nexus as an approach that supports integrative modelling of trade-offs within socio-ecological systems with the objective of informing decisions relating to management of environmental resources, delivery of public services and associated risks and with the potential to impact upon water, energy and food security and planetary boundaries" (p.3).[27]

[17]   Stephanie Galaitsi, Jason Veysey and Annette Huber-Lee. 2018. Where is the added value? A review of the water-energy-food nexus literature. SEI working paper.

[18]   Ozturk, I. 2015. Sustainability in the food-energy-water nexus: Evidence from BRICS (Brazil, the Russian Federation, India, China, and South Africa) countries. *Energy* 93: 999–1010.

[19]   Gulati, M., Jacobs, I., Jooste, A., Naidoo, D. and Fakir, S. 2013. The water energy-food security nexus: Challenges and opportunities for food security in South Africa. In at the Confluence—Selection from the 2012 World Water Week. J. Lundqvist (ed.). Vol. 1.150–64.

[20]   Lele, U., Klousia-Marquis, M. and Goswami, S. 2013. Good governance for food, water and energy security. *Aquatic Procedia* 1: 44–63.

[21]   Stephanie Galaitsi, Jason Veysey and Annette Huber-Lee. 2018. Where is the added value? A review of the water-energy-food nexus literature. SEI working paper.

[22]   Keulertz, M. and Woertz, E. 2015. Financial challenges of the nexus: pathways for investment in water, energy and agriculture in the Arab world. *International Journal of Water Resources Development* 31(3): 312–25. DOI:10.1080/07900627.2015.1019043.

[23]   Benson, D., Gain, A. and Rouillard, J. 2015. Water governance in a comparative perspective: From IWRM to a 'nexus' approach? *Water Alternatives* 8(1). http://www.idaea.csic.es/medspring/sites/default/files/Water-Governance-in-ComparativePerspective-From-IWRMto-a-Nexus-Approach.pdf.

[24]   Biba, S. 2015. The goals and reality of the water–food–energy security nexus: the case of China and its southern neighbours. *Third World Quarterly* 7(1): 51–70. DOI:10.1080/01436597.2015.1086634.

[25]   Bach, H., Bird, J., Jonch, C.T., Morck, J.K., Baadsgarde, L.R., Taylor, R., Viriyasakultorn, V. and Wolf, A. 2012. Transboundary River Basin Management: Addressing Water, Energy and Food Security. Mekong River Commission, Lao PDR. http://www.mrcmekong.org/assets/Uploads/M2R-report-address-water-energy-food-security.pdf.

[26]   Treemore-Spears, L.J., Grove, J.M., Harris, C.K., Lemke, L.D., Miller, C.J., Pothukuchi, K., Zhang, Y. and Zhang, Y.L. 2016. A workshop on transitioning cities at the food-energy-water nexus. *Journal of Environmental Studies and Sciences* 6(1): 90–103.

[27]   Mathew Kurian. 2018. Water-Energy-Food (WEF) Nexus Interactions and Agricultural Research for Development—the Case for Integrative Modelling via Place-based observatories. Background Papers. Science Forum 2018. Stellenbosch, South Africa. 10–12 October 2018.

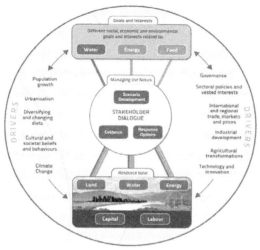

Figure 1.1 A conceptual diagram of the FAO approach to the Water-Energy-Food Nexus.

Departing from a static, water-centric interpretation of the Nexus, FAO has developed a conceptual approach to make sense of, and manage the complex and interlinked uses of water, energy and food (Fig. 1.1). The approach distinguishes between the resource base and the different goals and interests that are to be achieved with the same, but limited resources. It is about understanding and managing these different resource user goals and interests, while maintaining the integrity of ecosystems. For that, FAO has identified three areas of work through which it can contribute, namely by providing evidence, by developing scenarios, and by designing and appraising response options.

This model is discussed in more detail in Chapter 6 by Alfarra.

Many valid sectoral approaches exist for the development of sustainable food and agriculture and there continues to be a great need to find solutions within sectors, for example, by improving water use efficiency. At the same time, we need a shared vision and effective mechanisms to deal with cross-sectoral issues.

ROAD MAP FOR THE BOOK

In the first introductory section, this chapter is followed by "Water, Food, Energy and Environment Nexus", by Atif Kubursi discusses the multiple roles water plays in the economy, society. The environment increases the complexity of structuring prices that capture scarcity and the multiple values of water. Economics alone cannot deal with this complex commodity, resource and input. A far more appropriate framework is the nexus one that connects water to food security, energy, the environment and the economy. It is within this framework that an efficient and equitable value for water can be structured. The author concludes by stating that water is valuable, but it is not thicker than blood and an economic price can be set that embodies many of the competitive efficiency criteria in addition to other valuable societal, economic and environmental objectives. It is this nexus approach that governs the true worth of water. The last chapter in the introductory section, "The Urban Water-Energy-Climate Nexus and the Role of Demand Management": by Bears discusses transitions in managing natural resources in the face of climate change and natural resources scarcity. The chapter examines the concept of demand management and discusses its implementation in a series of mini case studies.

In the case study section, Yengoh explores, in "The Water-Energy-Food Nexus in Africa" the issues of inadequacy of access to water including social, economic, political, and even cultural. Other challenges such as climate change and land degradation do also contribute to this limited access and are forecast to become increasingly influential in the future. The WEF nexus provides a basis for understanding the interconnectedness of these three resource sectors and a framework for the development of policies and strategies for their sustainable management. Africa's rich natural resource base can form the foundation for development and management of WEF nexus components and for achieving sustainable green growth. The chapter explores the obstacles to implement nexus approach in Africa and some opportunities include developing adaptive governance that promotes flexibility, cooperation and collaboration to cope with the challenges of integrated development across WEF nexus sectors.

In Chapter 5, "The Water-Energy-Food Nexus in Europe", Fabiani et al. discuss the interdependence of water, energy and food security in the context of Europe and look at the European 2020 strategy for energy reduction, greenhouse gases reduction as well as the Water Framework Directive (2000/60/EC). The EU water policy has undergone a process of restructuring. The WFD is supplemented by international agreements and legislation relating to water quantity, quality and pollution. In Chapter 6 "The Water-Energy-Food Nexus in the Arab Region", Alfarra explores the FAO explanation of the water, energy and food nexus and apply it to the Middle East region. In the next chapter, Pebbles, in "Recognizing the Food-Energy-Water Nexus: A Renewed Call for Systems Thinking" gives a perspective of water, food, energy nexus perspective from North American perspective. In the last chapter of this section, Gabrielsson et al. in. "Linking the Water-Food-Energy Nexus to Sanitation: will it save and improve lives?" The authors discuss the global sanitation challenge. Since the problem is a social, technological, environmental, economic and political one, the authors explore how a nexus approach can help to broaden the dialogue so that areas of mutual benefit can be high-lighted to further the mission of safe and productive sanitation.

In the next section on "water and human rights", the first chapter by Grover gives a historical development of human rights, how this was extended to the environmental field and concludes with how human rights were finally linked to water. The next chapter by Alfarra et al. explore how enshrining these rights into the constitution make it easy for implementation and to interpret it for legal purposes.

In the next section on "Water security" the first chapter by Grover discusses water security as a new paradigm and challenges and opportunities under this paradigm. The next chapter by Alfarra redefines the concept of water security in light of the refugee crisis and the last chapter in this section illustrates this redefinition of water security and water-energy-food nexus by applying it to Jordan River Basin. In the last chapter, Humber and Bradbee explore the need and importance of education for implementing new policies and paradigms.

The Nexus approach offers an approach to go away from silos, towards coordination across sectors in decision making and it also offers a "framework to view policy and management interventions as outcomes of choices that operate at global, national and local scales, guided in turn by norms, agency and individual behavior with regards to allocation of financial and human resources and institutional capacity with the goal of balancing bio-physical risks with institutional ones that may subsequently be reflected in differing emphasis on advancing efficiency and equity considerations in infrastructure operation and maintenance" (p.3).[28]

---

[28]   Mathew Kurian. 2018. Water-Energy-Food (WEF) Nexus Interactions and Agricultural Research for Development—the Case for Integrative Modelling via Place-based observatories. Background Papers. Science Forum 2018. Stellenbosch, South Africa. 10–12 October 2018.

Chapter-2

Water, Food, Energy and the Environment Nexus

Atif Kubursi

INTRODUCTION

Is water different from other producible commodities? Is it an asset, a consumer good or a factor of production? If it is a resource, is it a renewable or a non-renewable resource? Is its value infinite and "thicker than blood?" Can a price for water be determined much like for any other commodity? Can water or water permits be traded? What constitutes an equitable distribution of this scarce resource? How does water relate to food security? How does water mix with oil and energy? What are the ecological demands for water? What is water's contribution to economic development?

These complex questions are provocative and can incite passionate debate, yet they are central to water resource management and for proper positioning of water within the food, energy, and environment nexus. In this chapter, we show that economics sheds useful light on how to address these questions. This is not to suggest that economics alone can answer all of them. There are complex economic, social, environmental and political issues that need to be factored into such an analysis. These add complexity to the general structure within which water will be considered, but do not change in any fundamental way the conclusions reached. They strongly suggest that the proper framework within which water issues can be dealt with is the nexus framework where the complex interrelationships of several systems are allowed to influence the analysis and discourse.

Before going any further in answering some of these questions it is useful to outline some of the specific attributes of water that distinguish it from other commodities and assets and set the stage for the nexus framework. These attributes

---

Emeritus Professor of Economics, McMaster University. Adjunct Professor, Faculty of Environmental Studies, York University.

will also influence the way water prices are set and the optimal allocation of this scarce resource. They also define the nature and complexity of the nexus that ties water to energy, food, the environment and the economy. These attributes and their consequences are listed below:

- Water is a fugitive, reusable, and stochastically supplied resource whose production can be subject to economies of scale which give rise to natural monopoly situations. In this respect water has many of the characteristics of a common property resource and a quasi public good.

- Water is typically a non-traded commodity that is rarely sold in a competitive market. There are few overt water markets where suppliers and demanders exchange water. Markets in water rights have emerged in several parts of the world; the most notable examples are in Australia, Colorado, California and Argentina (Saliba and Bush, 1987; Newlin et al., 2002; Grafton et al., 2010).

- Water values generally differ from the price that would be obtained in a free and competitive market. It is often the case that water has social value that is above what private users are willing to pay for it (Fisher, 2000). The allocation of water often reflects national and social policies and priorities towards agriculture, the environment, food security, energy security and national security that go beyond serving the interests of private farmers. Social and policy considerations apart, the diversion of actual prices from their scarcity values imposes social costs on the domestic economy as well as on neighbouring countries. Water in this perspective is a critical developmental vehicle with wide implications for growth and equity, Water is also part of the tragedy of the commons (Hardin, 1968).

- Water is a scarce resource whose availability is far below the competing demands for it. This scarcity is more pronounced in some specific regions and within even the same state. It is not uncommon to find areas and groups of people with abundant water and others with no or little amounts. The scarcity issue is about **relative** scarcity and not **absolute** scarcity.

- Water scarcity is also about **physical scarcity** that is complicated by **economic scarcity** where actual prices for water are fractions of its true scarcity price (shadow). When prices are below scarcity prices, waste and over use are quickly observed. There are many examples in the Mediterranean countries, particularly in Israel and Jordan where subsidies have engendered a culture of excessive use, irrational production structures and ultimately waste.

- The current allocation of shared water resources in the world are not the outcome of agreements, negotiations or equitable principles. Rather they reflect the asymmetries of power in existence and the abilities of the strong to impose their will on the weak. There is a deep and profound dichotomy between the balance of power governing current water allocations and the balance of interest of the riparian parties; a situation that cannot be sustained. Sooner or later there will arise some dynamic forces for restoring some symmetry between power and interest.

- Water is not only a desirable commodity, its availability is also critical for life. There are little or no substitutes for it. Furthermore, it is a well entrenched principle that no matter how scarce water is, every person is entitled to a minimum quantity that is considered consistent with human dignity (UNGA, 2010; Gleick, 1998).

- The secure supply of water in many regions of the world is quite low. Security of supply is defined here as the probability of its average flow availability 9 out of 10 years. In Canada, a country rich in water, this probability is less than 30% where as it is less than 5% in the Middle East (Lonregan and Brooks, 1994).

- While the total water supply may be limited and, few if any, substitutes exist for it; there exists nonetheless substantial possibilities for inter-sectoral and interregional substitutions. As well, there are a number of technologies and conservation programs that rationalize demand and raise the efficiency of its use in agriculture, industry and other sectors.

- Part of the world water scarcity crisis is the fact that agriculture on average uses about 70% of the total global supply (Anand, 2007). It is generally the case that other needs are suppressed, but this leaves a wide room for inter-sectoral reallocations and shuffling water between users and uses, spatially and temporally.

- Food security is ultimately tied to water security. Getting the highest value per drop of water is the main objective. Wasting food is wasting water. On average, 30 to 40% of food is wasted. This means that considerable amounts of water are wasted.

- While the quantity of water is in short supply, concern for preserving its quality is perhaps more pressing. Pollution and saline intrusion of the aquifers are being increasingly recognized as critical factors in planning for the future.

- Water is the "universal solvent" (Young, 2005). It performs numerous ecosystem services including the absorption, transportation, and dilution of pollutants. Water is indispensible to the environment's pollution abatement capacity.

- Water absorbs energy in its production, delivery and treatment. In some regions water is solely produced by energy through desalination. It is believed that in the next decade Arab Gulf oil producers may need to use half of their oil and gas energy supplies to simply meet the water demands of a growing population.

- Oil production plants consume large quantities of water and so do cooling systems. Every barrel of oil derived from the tar sand sources requires 10 barrels of water to separate the bitumen from sludge.

- One hundred and ten thousand cubic kilometres of precipitation, nearly 10 times the volume of Lake Superior falls from the sky as precipitation every year. This huge quantity would be enough to satisfy the requirements of everyone on the planet if this water falls where and when it is needed. About 56% of this water flows through the landscape and is never captured. It either evaporates or transpires from plants. Allen (2012) calls it green water to signify its illusiveness. Another 36% of the total available water channels through blue water sources— rivers, lakes, wetlands and aquifers—that people can tap directly.

- Farm irrigation is the largest single human use of freshwater. Cities and industries consume small fractions of the blue waters, but some of them draw more resources than is available at the local basin.

- Most countries 'import' embedded water. Over 160 countries of the world are net food importers and about 15% of food produced is traded internationally (Allen, 2012).

- Dependence on virtual water is risky. It takes a considerable amount of water to produce food. Actually, it takes 1000 tonnes (m^3) of water to produce a tonne of wheat and 16,000 tonnes of water to produce a tonne of beef.

- For virtual water to meet the needs of countries with large water deficits it is necessary to keep one billion Indians on vegetarian diets and we need to persuade 1.3 billion Chinese to prefer pork and chicken to beef (Allen, 2012).

Competitive Markets, Water Prices and Total Economic Value

Water is a scarce resource (asset), a scarce commodity and a scarce input. Economics is particularly suited for dealing with scarcity as economics after all is the study of how scarce resources are or should be allocated to various uses and users.

Indeed, water has a price as does any scarce resource, input or asset. There is a monetary equivalent to water although this is not the way people typically speak about water. This price need not emerge from competitive markets. But it has or could be constructed to have many of the characteristics that are associated with competitive prices. The price should reflect the scarcity of water and should disseminate information about this relative scarcity. It should also invite, if not provoke, the correct (incentives) responses that prices in general are expected to do. Higher prices persuade consumers to economize on its use and persuade suppliers to raise their quantities offered. In competitive markets, the market price is what buyers are willing to spend for an additional unit of the commodity and the extra cost of producing it (marginal cost). If the price is larger than the marginal cost of the last unit of water this provides a signal that more is needed of the commodity and more should be produced (Mansfiled and Yohe, 2003). Conversely, if the price is below marginal cost, additional units will not be produced as society will have to sacrifice more than what people would value the additional units.

It is generally accepted however, that water is not bought and sold in competitive markets. This is because in the case of water at least five of the basic properties of competitive markets are absent (Fisher, 2000). These five properties include the following:

First, free markets lead to an efficient allocation of scarce resources if these markets are characterized by competitive structures, that is, these markets include a large number of independent small sellers and a similarly large number of independent small buyers that no single supplier or buyer is significant enough to influence the price. Each and every buyer and seller in this market is a price-taker.

Second, competitive markets require freedom of entry and exits and that no barriers exist that preclude easy entrance or exit.

Third, the product must be homogeneous enough that each unit is quite similar to any other unit.

Fourth, for a free market to lead to an efficient allocation externalities must be absent. In economics, an externality or spillover of an economic transaction is an impact on a party that is not directly involved in the transaction. An efficient allocation can emerge from a free market when social costs coincide with private costs. Water production, however, involves many "externalities". In particular, extraction of water in one place reduces the amount available in another. Further, pumping water from an aquifer in one location can affect the cost of pumping elsewhere. Such externalities do not typically enter the private calculations of individual producers and drives a wedge between private cost and social costs.

Fifth, in a free market that allocates efficiently scarce resources social benefits must coincide with private ones. If not, then (as in the case of cost externalities) the pursuit of private ends will not lead to socially optimal results. In the case of water, many uses have social benefits that exceed the private ones. The use of water in agriculture may result in benefits that exceed the private returns to farmers. Among these are food security, border security, and national interest.

These conditions are often violated in the case of water, where water sources are relatively few, barriers to entry are real and high (high cost of infrastructure), a large gap exists between private and social costs and benefits and water units are not homogeneous with a large spectrum of different qualities are observed. This is why perhaps that it is often the case that water production facilities are owned by the State. In many respects water is not a private good; it has what we alluded to above, many of the characteristics of quasi-public goods.

The fact that water is not bought or sold in competitive markets does not mean it does not have a price (Eckstein, 1958; Maass et al., 1962; Hirshleifer et al., 1960). It simply means that this price is not the one where competitive conditions prevail. It is possible, however, to build specific models of water allocations that simulate competitive conditions and where the optimal nature of markets is restored. The model can explicitly optimize the benefits to be obtained from water, taking into account the five points made above. It is also true that now we have a rich body of literature that informs the setting of water prices and tariffs to overcome some of the five points listed above. The prices that emerge from formal optimization models not only maximize profits for the producers (producer surplus), but could also maximize the utility of the consumer (consumer surplus). Furthermore, as is shown later, a multi-objective framework can be developed to design prices that satisfy goals beyond pure economic efficiency. These constructed prices permit the optimal allocation of water to its best uses as would emerge under competitive conditions. In many respects, the constructed prices are designed to serve as guides to consumers and producers in much the same way as competitive prices.

The Optimizing Model: Simulating a Competitive Water Market

The underlying economic theory of the "optimizing model" is simple and compelling.

As already discussed, purely private markets and the prices they generate cannot be expected to optimize net social benefits of water. Nevertheless, prices in an optimizing model can play a role very similar to that which they play in a system of competitive markets. Typically these models allocate water so as to maximize the net benefit (consumer surplus plus producer surplus) obtained from this allocation. The maximization of net benefits is done subject to constraints. An example of such constraints includes that at each location; the amount of water consumed cannot exceed the amount produced there plus net imports (imports minus exports) into that location (Amir and Fisher, 1999; Fisher and Huber-Lee, 2008).

Corresponding to the optimum quantities, there is a dual solution that defines the associated prices at the optimized level of activities. It is a general theorem that when maximization involves one or more constraints, there is a system of prices involved in the solution. These prices are called "shadow prices" or "Lagrange multipliers". Each shadow price shows the rate at which the quantity being maximized would increase if the associated constraint were relaxed by one unit at the optimal solution. In effect, the shadow price is the amount the maximizer would be just willing to pay (in terms of the quantity being maximized) to obtain an additional unit of the associated constrained quantity.

Derivation of shadow prices

The calculation of a shadow price of water is sketched in a simple example below by way of illustrating its derivation and interpretation. We start with postulating the following optimizing problem:
Minimize cost of production:

$$\bar{w}L + \bar{e}K + \bar{r}W$$

Where
w = wages
e = rental cost of capital
r = rental price of water
L = labour
K = capital
W = water

The bar over the prices is meant to denote competitively determined prices. Subject to an aggregate production functional relationship

$$\bar{X} = F(L, K, W)$$

And a fixed amount of water

$$W \leq \bar{W}$$

Where $\bar{X}$ is the fixed total output

The Lagrangian

$$£(L,K,W,\lambda_1,\lambda_2) = \overline{w}L + \overline{e}K + \overline{r}W + \lambda_1\left[\overline{X} - F(L,K,W)\right] + \lambda_2\left(\overline{W} - W\right)$$

Differentiate totally: d£ =

$$\overline{w}dL + \overline{e}dK + \overline{r}dW + \lambda_1\left[-\frac{\partial F}{\partial L}dL - \frac{\partial F}{\partial K}dK - \frac{\partial F}{\partial W}dW\right] + \lambda_2\left[d\overline{W} - dW\right] + \left[\overline{X} - F(L,K,W)\right]d\lambda_1 + \left[\overline{W} - W\right]d\lambda_2$$

Group terms: d£ =

$$\left(\overline{w} - \lambda_1\frac{\partial F}{\partial L}\right)dL + \left(\overline{e} - \lambda_1\frac{\partial F}{\partial K}\right)dK + \left(\overline{r} - \lambda_1\frac{\partial F}{\partial W} - \lambda_2\right)dW + \lambda_1d\overline{X} + \lambda_2d\overline{W} + \left[\overline{X} - F(L,K,W)\right]d\lambda_1 + \left[\overline{W} - W\right]d\lambda_2$$

The first three terms are the first order conditions of minimizing costs and are set equal to zero.

$$\frac{d£}{d\overline{X}} = \lambda_1$$

Where, λ_1 is the influence on cost of producing one additional unit of output or the marginal cost (MC) of output.

$$\frac{d£}{d\overline{W}} = \lambda_2$$

Where, λ_2 is the shadow price of water as it depicts MC of one additional unit of water.

The first order condition for water implies the following:

$$\overline{r} = \lambda_1\frac{\partial F}{\partial W} + \lambda_2 \qquad\qquad \lambda_1, \lambda_2 \geq 0$$

but λ_1 = price of output under optimal competitive conditions therefore

$$\overline{r} = P\frac{\partial F}{\partial W} + \lambda_2$$

Therefore the rental price of water would equal its value of a marginal product plus a scarcity premium represented by λ_2.

The shadow price is simply a Lagrangian multiplier in the optimization equation. It denotes the improvement in the Objective Function due to a relaxation of any given constraint.

Shadow prices generalize the role of market prices as can be seen from the following propositions:

- Where there are only private values involved, at each location, the shadow value of water is the price at which buyers of water would be just willing to buy and sellers of water just willing to sell an additional unit of water.

- Where social values do not coincide with private values, this need does not hold. In particular, the shadow price of water at a given location is the price at which the user of water would be just willing to buy or sell an additional unit of water. That payment is calculated in terms of net benefits measured according to the user's own standards and values.

- Water *in situ* should be valued at its scarcity rent. That value is the price at which additional water is valued at any location at which it is used, less the direct marginal costs involved in transporting it there.

Structuring a price (Tariff) of water

Water utilities routinely design water tariffs and prices. Their choice of the price is motivated by different considerations than pure economics. They simply choose the price at a level that would help cover their operating costs plus a capital charge in addition to several other objectives. Few utilities, if any, seek solely to generate sufficient revenue to cover their private costs. The generation of revenue is rarely the only purpose of a tariff, nor is it the sole consideration (Boland, 1993). It has been generally recognized that water tariffs are powerful management tools with a number of complex and important functions. The tariff could create incentives for the efficient production and use of water. In this way the tariff promotes environmental sustainability. As well, tariffs can be set to recover the full cost (including externalities, both economic and environmental)[1] of an activity or structured to subsidize poorer customers. Each objective would lead to a different tariff design. The "best" design is the "one that strikes the most desirable balance among the different objectives" (Boland, 1993). But is the "best" tariff the optimal one? Standard optimization theory tells us that net social benefits are maximized where marginal social benefit equals marginal social cost. But there is more than one cost perspective to consider when structuring an optimal tariff or price for water. Given the many different definitions for full supply cost, full economic cost, and full cost, the question remains as to which marginal cost should be equated with the marginal benefit. Full cost pricing requires a quantitative assessment of water's intrinsic and cultural values, as well as the net of environmental externalities.

The legitimacy of any such quantification is, however, severely limited by the subjectivity and unavailability of relevant data (Rogers et al., 1998; Conca, 2006; NRC, 2005). But optimality requires quantification and this has become a major issue in setting the optimal water tariff. The different definitions and their relevance for designing the optimal water tariff are presented below.

Opportunity cost

The opportunity cost of a commodity or service is the value of its next best alternative use. If good X has only two possible uses, A and B, the opportunity cost of using good X for use A is the foregone benefit of employing it in use B. Similarly, the opportunity cost of using X for B is A. That is, it is the value of the foregone best alternative use. As Rogers et al. (1998) argue, neglecting the value of water's opportunity costs "undervalues water, leads to failures to invest, and causes serious misallocation of the resource between users."

---

[1]   Rogers, Bhatia, and Huber (1998) make an interesting distinction between economic and environmental externalities. Economic externalities are more readily quantifiable and involve costs such as those imposed on downstream users by upstream diverters and polluters. Environmental externalities include the value of ecosystem degradation, the loss of biodiversity, and the health effects on human, plant, and animal populations.

Full supply cost

This is the sum of costs directly involved in delivering water to the end user. This may include pumping, purification, storage, fixed infrastructure costs, capital investment costs, government taxes, and connection and metering costs. Full supply costs exclude all externalities and opportunity costs (Rogers et al., 1998). Capital and infrastructure costs are particularly heterogeneous vectors that vary widely depending on the types of delivery systems involved. Wells, house connections, pipes of various flow capacities, desalination projects, and recycled and treated water can lead to vastly different nominal values for full supply cost (Perks and Kealey, 2006).

Full economic cost

This is an economic concept that refers to the full supply cost plus the opportunity costs and net of economic externalities.

Full cost

This is the full economic cost plus environmental externalities (these are more diffused, indirect and not easily quantifiable) (Fig. 2.1).

The following principles and objectives have been identified as necessary ingredients in determining a price for water. Each of these is broadly defined and adopted in varying degrees with different weights placed on each from case to case.

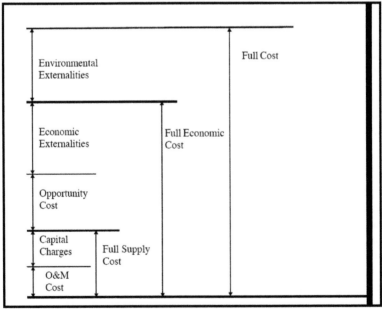

Source: Rogers et al. (1998).

Figure 2.1 Full cost prices.

In Fig. 2.2 below we sketch a multiple criteria framework for choosing prices that deal with complex objectives that balance economic, environmental and socio-cultural issues. In other words, we conceptualize the existence of a deterministic nexus that guides water values and allocations to achieve multiple objectives within a dynamic interactive system that balances multiple and competing ends.

Even the pursuit of economic efficiency incorporates multiple goals. Fees for water services must be sufficient to ensure that the utility recovers at least the full supply cost of delivery to the end user, a concept known as revenue sufficiency. Any fraction less than this fails to send the signal that water is a scarce resource, withholds information from consumers, the utility, and the regulatory agency, and prevents the proper inter-agent allocation of water.

It is worth noting that the standard approach to economic efficiency includes the concept of 'full cost recovery,' which the United Nations World Water Development Report defines as a situation in which "users pay the full cost of obtaining, collecting, treating, and distributing water, as well as collecting, treating, and disposing of wastewater" (WHO and UNICEF, 2010). In practice, as well as in theory, 'full cost recovery' and 'revenue sufficiency' refer only to the full supply cost (Boland and Whittington, 2000; Whittington, 2003; Yepes, 1999; Boland, 1993; Rogers et al., 1998). We argue below that this is undesirable. This particular characterization is, however, of interest because it directly addresses the management of wastewater. That the importance of the economics of wastewater management has been largely ignored is demonstrated by the vast differences in progress towards the achievement of the Millennium Development Goals for water and sanitation (WHO and UNICEF, 2010).

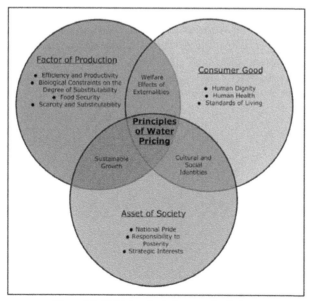

Figure 2.2 Multiple criteria framework of prices.

In stark contrast to convention, we argue that pricing at the full cost of supply does not support true economic efficiency. Supply cost pricing brings us closer to the outcome that would exist if water were traded in a perfectly competitive market. This price for water would fail to properly incorporate the value of opportunity costs, economic externalities, and environmental externalities (Rogers et al., 1998; Rogers et al., 2002). Any such price prevents socially optimal allocations by promoting over use, undervaluation, and waste. For this reason we argue that wherever possible full cost or full economic cost, as they are defined above, should be used in the development of water tariffs. These prices reflect the nexus framework of multiple accounting structures embedded in the nexus approach.

Water Tariffs, Pricing Strategies and Fairness

The water tariff is the primary tool with which a policy maker can combine the various objectives and considerations into a socially optimal price regime. The tariff is central to the utility's ability to attract capital, creates incentives for efficiency in production and consumption, and distributes costs across consumers and time (Boland, 1993).

Traditionally, Decreasing Block Tariff (DBT) pricing strategies have been employed in water delivery systems for several reasons (Griffin, 2006). The decreasing block tariff, as shown in Fig. 2.3 involves prices based on quantity consumed.

The initial block, or quantity of water, costs the most and prices decrease as consumption rises. DBTs score well on stability grounds as revenue is insulated from climatic shocks (to both supply and demand) and because the largest share of revenue comes from the initial units of water consumed. They exploit the economies of scale that arise from falling average costs as quantities supplied rise (Griffin, 2006). The problem, however, is that on grounds of efficiency, equity, and sustainability, this structure has less to offer than an increasing block tariff scheme. In terms of efficiency, DBTs send exactly the wrong message to consumers. They fail to convey the fact that water is a scarce resource that must be conserved; they promote waste,

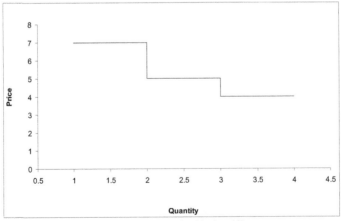

Figure 2.3 Decreasing block tariffs.

and fail to encourage socially optimal inter-agent, inter-sectoral, and inter-temporal allocations. Furthermore, as the utility reaches capacity, the marginal cost of supply it faces rises, however, under the DBT regime, the marginal cost faced by consumers is falling. In terms of equity and fairness, the DBT fails to treat equals similarly because consumers of water with identical costs of supply face different prices. By charging higher prices for the initial units of water, the DBT also ensures that the poor who consume the least water face the highest prices.

In recent years, DBTs have largely been replaced by Increasing Block Tariffs (IBTs), which have become the strategy of choice in the developing world. This is because they have gained preference by donor institutions, management, engineering, and financial advisors, and water experts as being the most effective way of synthesizing the goals and principles of water supply and demand (Boland and Whittington, 2000). The IBT (the pricing scheme depicted in Fig. 2.1) also features different prices based on quantities consumed, however, in this case prices rise with quantity consumed.

In terms of economic efficiency, this is a significant improvement over the DBT. Here, tariffs are consistent with marginal cost pricing and create an incentive structure that conveys the necessary signals to encourage socially optimal use. IBTs score well on fairness grounds because income and water demand are positively correlated and the prices faced by consumers rise with consumption. That is, IBTs are progressive, so wealthier customers that consume more water face higher prices. Furthermore, IBTs can exhibit 'internal subsidization' or 'cross subsidization' in which large consumers pay sufficiently high prices that the first units of water consumed can be sold below cost. This feature enables the utility to provide water to the poor below the full cost of supply without violating the principle of revenue sufficiency and other environmental concerns.

The practical implementation of IBTs has yielded mixed results. If blocks are not properly structured in quantities that are relevant for the consumers in question, the benefits of IBTs are lost (Olmstead and Stavins, 2007; Rogers et al., 2002; Whittington, 2003). Specifically, the compatibility of fairness with the other principles of water pricing requires that the first block be sufficient to cover the basic requirements of human health and dignity, but not more. More attention must be paid to determining the relative block sizes and their prices and finding the appropriate values for both requires data that is frequently unavailable.

It is possible that in some cases low initial block prices may yield a result in which metering costs may exceed the revenue collected from low-income users. In such situations, the utility provider incurs a loss not only from the provision of water below full supply cost, but on the metering operations as well (Yepes, 1999). Alternatively, if tariffs faced by large-scale consumers such as heavy industry and agriculture are sufficiently high, these consumers may opt to disconnect from the utility and seek alternative sources of water such as in house direct extraction (Whittington, 2003; Yepes, 1999). Even where this is illegal, government capacity to enforce such laws is frequently limited. The effect of this would be to remove the highest paying customers, which in turn would eliminate the feasibility of internal subsidization, increase average costs, and distribute fixed capital costs across a smaller customer base. That is, the benefits of IBTs are highly dependent upon the participation of the highest paying customers; however, these are exactly the customers who have the greatest incentive to search for alternative sources of water.

Even where subsidies are absolutely necessary in order for the poor to meet their basic needs, it may not be ideal to provide these subsidies through the tariff structure. Chile subsidizes water consumption for the poor via transfer payments from municipal and national budgets. Many have claimed that this approach enhances equity and economic efficiency because the tariff received by the utility 'treats equals as equals' (Yepes, 1999). Although we accept Yepes' (1999) claim in principle, we do so with some trepidation. For the argument to hold, it must be shown that the welfare loss incurred by using the tariff structure to internally subsidize water for the poor is greater than that incurred by appropriating funds from elsewhere in the economy to fund government transfers. This reservation may be especially relevant in developing countries where the tax base is small, government capacity is limited, and corruption is significant.

An interesting extension to the principle of IBTs involves the concept of two-tariff design. The price that end users face for water can be divided into two components: volumetric usage charges and fixed rate charges commonly referred to as connection fees. The two-tariff system provides an economic incentive for end users to conserve water through its volumetric component while providing an element of revenue security for the supplier through its fixed rate component. This structure has evolved into the structure of choice in most OECD countries (Griffin, 2006).

A two-tariff system can be carried out in the context of an IBT scheme so that a fixed cost is paid for 'connection' but low levels of water can still be provided at a rate that is subsidized by larger consumers. In many areas, connection fees are sufficiently high that they exclude the poorest consumers. Since the cross-subsidization inherent in the IBT structure is only beneficial to those with connections, an argument can be made for higher subsidies on connection costs combined with lower subsidies on volumetric consumption. This approach incorporates the poorest strata of society while continuing to expand the customer base, which decreases average costs for all.

This is especially important in situations where land tenure rights are insecure. In such cases the incentives and capacity for household investment in connections to the water utility are extremely low. It may be the case that these areas realize their greatest economic surplus by providing shared public taps that face a fixed rate rather than an IBT. There is already some evidence to support such a conclusion. Poor neighbourhoods that face IBT prices often share a single connection between multiple households. These meters read high volumes of consumption and so the absolute poorest face higher prices due to the structure of the IBT.

Raising the price of water to its full cost would enhance economic efficiency and sustainability. Furthermore, Rogers et al. (2002) argue that this can be done without sacrificing equity and fairness. Situations exist in which Pareto improvements (improvements that leave at least some agents better off without making others any worse off) are possible even in cases where relative equity decreases (Whittington, 2003).

A nexus approach calls for integrating multiple objectives in the choice of the price regime. The discussion above is motivated by such considerations as it demonstrates that this integrative approach is not only feasible but necessary for ensuring environmental sustainability, development and equity.

Conclusion

Water is not traded in competitive markets. High infrastructure costs raise barriers to entry ruling out competitive conditions whereas economies of scale, limited substitutes, human rights issues and environmental externalities combine to complicate the design of a fair, efficient and sufficient price for water.

The multiple roles which water plays in the economy, society and the environment increase the complexity of structuring prices that capture scarcity and the multiple values of water. Economics alone cannot deal with this complex commodity, resource and input. A far more appropriate framework is the nexus one that connects water to food security, energy, the environment and the economy. It is within this framework that an efficient and equitable value for water can be structured.

Optimal allocations of this scarce resource among competing users and uses are not solely economic issues. The optimality criteria could be salvaged and applied where net social benefits are maximized within an idealized model capable of simulating competitive conditions but taking into account the relationships of water to agriculture, energy and equity. The shadow prices for water that emerge from the model solutions can be used to guide allocation of this scarce resource to optimize net social benefits inclusive of non-economic considerations. It is also feasible to construct a price designing exercise where it is possible to capture the full cost of producing and delivering water and to preserve and balance multiple objectives including environmental and equity considerations. Many utilities do that now on a routine basis.

Water is valuable but it is not thicker than blood and an economic price can be set that embodies many of the competitive efficiency criteria in addition to other valuable societal, economic and environmental objectives. It is this nexus approach that governs the true worth of water.

References

Allen, T. 2012. Communicating the underlying fundamentals of water security & the challenge of managing ignorance: how the concept of virtual water helps. Presentation at London School of Economics. January, 20.

Amir, I. and Fisher, F.M. 1999. Analyzing agricultural demand for water with an optimizing model. *Agricultural Systems* 61: 45–56.

Anand, P.B. 2007. Scarcity, Entitlements and the Economics of Water in Developing Countries. Northampton, MA: Edward Elgar Publishing Ltd.

Boland, J. 1993. Pricing urban water: principles and compromises. *Water Resources Update: Universities Council on Water Resources* 92(Summer): 7–10.

Boland, J.J. and Whittington, D. 2000. The political economy of increasing block tariffs in developing countries. pp. 215–235. *In*: Dinar, A. (ed.). The Political Economy of Water Pricing Reforms. Oxford, UK: Oxford University Press.

Conca, C. 2006. Governing Water: Contentious Transnational Politics and Global Institution Building. Cambridge, MA: MIT Press.

Eckstein, O. 1958. Water Resources Development: The Economics of Project Evaluation. Cambridge, MA: Harvard University Press.

Fisher, F.M. 2000. Towards cooperation in water: The Middle East water project. *Revue Région et Développement* 12: 143–165.

Fisher, F.M. and Huber-Lee, A. 2006. Economics, water management, and conflict resolution in the Middle East and beyond. *Environment* 48(3): 26–41.

Fisher, F.M. and Huber-Lee, A.T. 2008. WAS-guided cooperation in water management: coalitions and gains. pp. 181–208. *In*: Dinar, A., Albiac, J. and Sánchez-Soriano, J. (eds.). Game Theory and Policymaking in Natural Resources and the Environment. New York: Routledge.

Gleick, P.H. 1998. The human right to water. *Water Policy* 1(5): 487–503.

Grafton, R.Q., Libecap, G.D., Edwards, E.C., O'Brien, R.J. and Landry, C. 2010. Water Scarcity and Water Markets: A Comparison of Institutions and Practices in the Murray-Darling Basin of Australia and the Western US. International Center for Economic Research. Working Paper No. 28.

Griffin, R. 2006. Water pricing. *In*: Water Resource Economics: The Analysis of Scarcity, Policies, and Projects. Cambridge, MA: MIT Press.

Hardin, G. 1968. The tragedy of the commons. *Science* 162: 1243–1248.

Hirshleifer, J., DeHaven, J.C. and Milliman, J.W. 1960. Water Supply: Economics, Technology, and Policy. Chicago, IL: University of Chicago Press.

Krautkraemer, J. 2005. Economics of Natural Resource Scarcity: The State of the Debate. Resources for the Future, Washington, D.C., April, Discussion Paper 05–14.

Lonergan, S.C. and Brooks, D.B. 1994. Watershed: The Role of Fresh Water in the Israeli-Palestinian Conflict, Ottawa: IDRC.

Maass, A., Hufschmidt, M.M., Dorfman, R., Thomas, H.A. and Marglin, S. 1962. Design of Water Resource Systems. Cambridge, MA: Harvard University Press: Cambridge.

Mansfield, E. and Yohe, G. 2003. Microeconomics: Theory and Applications, 11. New York: W. W. Norton & Company.

National Research Council. 2005. The meaning of value and use of economic valuation in the environmental policy decision-making process. *In*: Mark Gibson (powerpoint presentation). Valuing Ecosystem Services: Toward Better Environmental Decision-Making. Washington, DC: The National Academies Press.

Newlin, B.D., Jenkins, M.W., Lund, J.R. and Howitt, R.E. 2002. Southern California water markets: Potential and limitations. *Journal of Water Resources Planning and Management* 128(1): 21–32.

Olmstead, S.M. and Stavins, R.N. 2007. Managing Water Demand: Price vs. Non-Price Conservation Programs, No. 39. Massachusetts: Pioneer Institute, 1–40.

Perks, A.R. and Kealey, T. 2006. International price of water. pp. 147–156. *In*: Aravossis, K., Brebbia, C.A., Kakaras, E. and Kungolos, A.G. (eds.). Environmental Economics and Investment Assessment. Boston, MA: MIT Press.

Rogers, P., Bhatia, R. and Huber, A. 1998. Water as a social and economic good: How to put the principle into practice. Global Water Partnership/Swedish International Development Agency. Stockholm, Sweden.

Rogers, P., de Silva, R. and Bhatia, R. 2002. Water is an economic good: How to use prices to promote equity, efficiency, and sustainability. *Water Policy* 4(1): 1–17.

Saliba, B.C. and Bush, D.B. 1987. Water Markets in Theory and Practice: Market Transfers, Water Values, and Public Policy. Boulder, CO: Westview Press.

Solow, R. 1994. An almost practical step toward sustainability. Resources Policy, 1993, MA, Washington, DC: National Academy Press 19(3): 162–172.

Turner, R.K., Pearce, D. and Bateman, I. 1993. Environmental Economics. Baltimore, MD: The Johns Hopkins University Press.

United Nations General Assembly. 2010. The Human Right to Water and Sanitation, A/Res/64/292, 1–3.

Whittington, D., Pattanayak, S.K., Yang, J.C. and Kumar, K.C. 2002. Do households want improved piped water services? Evidence from Nepal. *Water Policy* 4(6): 531–556.

Whittington, D. 2003. Municipal water pricing and tariff design: a reform agenda for South Asia. *Water Policy* 5(1): 61–76.

World Health Organization and UNICEF. 2010. Progress on Sanitation and Drinking Water: 2010 Update, Geneva: WHO Press.

Yepes, G. 1999. Do Cross-Subsidies Help the Poor to Benefit from Water and Wastewater Services? Lessons from Guayaquil, UNDP-World Bank Water and Sanitation Program: Washington, D.C., 1–9.

Young, R.A. 2005. Water, economics, and the nature of water policy issues. pp. 3–16. *In*: Resources for the Future (ed.). Determining the Economic Value of Water: Concepts and Methods. Washington, D.C.

Zerah, M.H. 2000. Water: Unreliable Supply in Delhi. New Delhi, India: Centre de Sciences Humaines.

Chapter-3

The Urban Water-Energy-Climate Nexus and the Role of Demand Management

Robert C. Brears

INTRODUCTION

By 2030, 5 billion people will be living in urban areas with hundreds of millions living in one of the world's 41 mega-cities, up from 28 today (United Nations Population Fund, 2016; United Nations, 2014). At the same time global demand for water is projected to exceed supply by 40% in 2030 and 55% in 2050 as a result of numerous trends including rapid urbanisation, water-energy nexus pressures and climate change (OECD, 2012; UNEP, 2016).

Already, cities around the world are facing water scarcity: India is projected to add 404 million urban dwellers between 2014 and 2050, however, government data revealed that residents in 22 out of 32 major cities already deal with daily water shortages (Times of India, 2013; UN, 2014). Meanwhile, the Middle East and North Africa (MENA) region is one of the most rapidly urbanising regions in the world with urban populations growing at 2.4% per annum (World Bank, 2014), yet the MENA region is one of the most water-scarce in the world: in 1950, per capita renewable water resources were 4,000 cubic metres per year and this has dropped to 1,100 metres in 2007 and will likely drop further to 550 by 2050 (World Bank, 2007).

The International Energy Agency projects that energy-related water consumption will increase by nearly 60% between 2014 and 2040: In the Asia-Pacific region, 3 out of 4 countries already face water scarcity, yet the region will see a rapid rise in energy consumption from around a third of global consumption to 51–56% by 2025 (ADB, 2013; UN-Water, 2014). Meanwhile, the amount of energy used in the water sector

Po Box 79032, Avonhead, Christchurch, New Zealand.
Email: rcb.chc@gmail.com

globally will more than double, mainly due to desalination projects, large-scale water transfer projects as well as increasing demand for wastewater treatment (IEA, 2016).

Climate change will lead to fluctuations in precipitation, ground and surface water levels as well as water quality, for instance, the FAO predicts that for each 1°C of global warming 7% of the world's population will see a decrease of 20% or more in renewable water resources (FAO, 2016). For example, in the Danube River Basin climate change will result in seasonal precipitation changes with a decrease in summer and an increase in winter, furthermore, an increase in water temperature will lower water quality (ICPDR, 2016).

With cities likely to face water scarcity from urban population growth, rising water-energy nexus pressures and climate change in the coming decades, urban centres will need to transition towards a demand management framework that balances rising demand with limited and often variable supplies of water.

This chapter will first discuss transitions in managing natural resources in the face of climate change and natural resources scarcity. The chapter will then discuss transitions in water resources management before introducing the concept of demand management before finally discussing its implementation in a series of mini case studies.

Transitions in Managing Climate Change and Natural Resources Scarcity

In transitions towards managing natural resources sustainably, there are two types of drivers to consider: climatic and non-climatic. Regarding climate change, there are two approaches society can take in adapting to the pressures of climate change: mitigation and adaptation. Traditionally, it is common for policymakers to mitigate the impacts of climate change by taking actions that prevent the impact of an event, for example constructing dams and reservoirs to protect communities from variable levels of precipitation. However, these 'hard' infrastructural solutions are typically economically and environmentally costly to implement (Australian Government Productivity Commission, 2012). In contrast, adaptation towards climate change aims to first increase the capacity of a system to successfully respond to climate change through behavioural, resource and technological adjustments and second, reduce the risks associated with the impacts of climate change (Adger et al., 2007; Kolikow et al., 2012).

There are two main types of adaptations in climate change: green actions and soft actions. Green actions ensure that ecosystem health is maintained in order to reduce society's vulnerability to risks; this can be achieved by ensuring natural resources are used as efficiently as possible. An example of green actions is constructing wetlands and restoring riverways to manage variable levels of precipitation. Green actions are usually less resource-intensive than mitigation (hard actions) in terms of financial and technological capacity, as green actions do not usually require the development and maintenance of high-tech, innovative solutions (European Environment Agency, 2013). In addition, green actions are also less environmentally costly to implement compared to mitigation, as they focus on preserving the health of ecosystems

(Australian Government Productivity Commission, 2012). Nonetheless, green actions frequently overlook the social dimensions of climate change; instead they focus on economic and technological solutions to the problems (Hoffman, 2010). In soft actions, the focus is on using management, legal and policy approaches to alter human behaviour as a way of reducing vulnerability to climate change risk (European Environment Agency, 2013).

Regarding natural resources scarcity due to non-climatic drivers, institutions typically rely on meeting actual or perceived supply inadequacies by constructing large-scale infrastructural projects to increase supply. However, because of the large environmental and economic costs associated with supply-side projects, natural resources managers have turned to policies which focus on improving economic and technological efficiency in managing the demand and supply of natural resources (Wolfe and Brooks, 2003). In this context, natural resources managers use economic and technological measures to manage natural resources more efficiently.

Nonetheless, while economic instruments and technological developments may appear to provide solutions to resource scarcity, it is individual beliefs, norms and values that drive resource scarcity (Hoffman, 2010; Lieberherr-Gardiol, 2008). As such, to properly address resource scarcity there needs to be a transition in societal values, in particular, changes in behavioural patterns, thinking and value structures regarding the environment so that society recognises that resource scarcity and environmental degradation is not only a scientific fact but a social fact too (Milbrath, 1995; Hoffman, 2010; Wolfe and Brooks, 2003). Therefore, the focus is on behavioural change as a way of decreasing demand for resources, which in turn lowers environmental degradation (Williams and Millington, 2004; Wolfe and Brooks, 2003). In this context, natural resources managers recognise that a transition is not only a technological matter but a social matter too, requiring deep and broad social relearning of thinking, value structures, behavioural patterns and institutional arrangements concerning scarce resources (Milbrath, 1995).

Transitions in Water Resources Management

In traditional water resources management, urban water managers, facing increased demand and variable levels of supply, have relied on large-scale, supply-side infrastructural projects such as dams, reservoirs and pipelines to meet increased demand for water (supply-side management) (Gleick, 1998; Sofoulis, 2005; Richter et al., 2013; Molle and Berkoff, 2009). Over time, however, these supply-side solutions have become unfavourable due to their environmental, economic and even political costs. Environmental costs include disruptions to waterways that support aquatic ecosystems, adversely impacting the numerous ecosystem services on which both humans and nature rely on (Molle and Berkoff, 2009). Economic costs stem from a reliance on more distant water supplies, often of interior quality, which not only increases the cost of transportation (energy) but also the cost of treatment (chemical) (Van Roon, 2007; Bithas, 2008). Politically, since the vast majority of water resources are transboundary, supply-side projects can create political tensions because they rely on water crossing both intra- and inter-state administrative and political boundaries (Brears, 2016). In addition, supply-side solutions fail to account for uncertainty in supply from climate change extremes (floods and droughts) and changing weather patterns (spatial and temporal changes in precipitation levels) (Molle and Berkoff, 2009; Van der Brugge and Van Raak, 2007). As such, with increased demand for water and variability of supply, supply-side solutions have become outdated (Bahri, 2012). Therefore, there is a need to transit towards managing actual demand for water (demand management), as it is peoples' attitudes and behaviour towards water that determines the amount of water needed (Brears, 2016).

Demand Management

Demand management involves better use of existing water supplies before plans are made to further increase supply. In particular, demand management promotes water conservation, during times of both normal conditions and uncertainty, through changes in practices, cultures and people's attitudes towards water resources (Savenije and Van Der Zaag, 2002). In addition to the environmental benefits of preserving ecosystems and their habitats, demand management is cost-effective compared to supply-side management because it allows the better allocation of scarce financial resources, which would otherwise be required to build expensive dams, water transfer schemes from one river basin to another, and desalination plants (Global Water Partnership, 2012). Urban water managers can implement demand management techniques to:

- Reduce loss and misuse in various water sectors (intra-sector efficiency).
- Optimise water use by ensuring reasonable allocation between various users (cross-sectoral efficiency) while taking into account the supply needs of downstream ecosystems and other water users and uses.
- Facilitate major financial and infrastructural savings for cities by Minimising the need to meet increasing demand with new water supplies.
- Reduce the stress on water resources by reducing or halting unsustainable exploitation of water resource.
- Achieve additional benefits including reducing energy demand and greenhouse gas emissions.

Demand Management Strategies

Demand management involves communicating ideas, norms and innovations for water conservation across individuals and society, the purpose being to change people's culture, attitudes and practices towards urban water resources and reduce consumption patterns (Global Water Partnership, 2012; Muller, 2007). As such, urban water managers can use two types of demand management strategies to modify attitudes and behaviour towards water: consequential strategies attempt to influence the determinants of target response after the performance of the behaviour. This assumes that feedback, both positive and negative, of the consequences of that behaviour, will influence the likelihood of the behaviour happening or not happening in the future; and antecedent strategies attempt to influence the determinants of target behaviour prior to the performance of the behaviour (Maheepala et al., 2010; Molle and Berkoff, 2009; Gifford et al., 2011). Using these two strategies, urban water managers can use two types of demand management instruments to achieve urban water security: (i) regulatory and economic instruments and (ii) communication and information instruments (Brears, 2016).

Regulatory and Economic Instruments

Regulatory instruments are frequently used in the management of water and involve setting allocation and water-use limits. In addition, economic instruments are used to provide incentives for all water users to conserve water and use it efficiently.

Conservation ordinances

Temporary urban water conservation ordinances and regulations restrict certain types of water use during specified times and/or restrict the level of water use to a specified amount. These programmes are usually enacted during times of severe water shortages and cease once the shortage has passed (Michelsen et al., 1999; Canada West Foundation, 2004). Examples of temporary water-use regulations include: (i) restrictions on non-essential water uses, e.g., watering lawns, washing cars, filling swimming pools, washing driveways; (ii) restrictions on commercial use, e.g., car washes, hotels and other large consumers of water; and (iii) bans on using water of drinking quality for cooling purposes (Brears, 2014).

Permanent urban water conservation ordinances and regulations include amendments to building codes or ordinances requiring water management plans or the installation of water meters and water-saving devices, e.g., low-flow toilets, showerheads and faucets in all newly constructed or renovated homes and offices (Michelsen et al., 1999; Pennsylvania State University, 2010). For example, plumbing codes can be used to ensure new homes and offices meet the maximum water-use standards for plumbing fixtures such as toilets, urinals, faucets and showers.

Case study 1: Western Australia's mandatory water efficiency management plans

Western Australia's Water Corporation has developed a drought-proof plan for Perth to ensure there is enough water for all, even in the face of climate change and an increasing population. The plan aims to achieve the city's goal of reducing water consumption by 25% by 2060. To help business and industrial customers reduce their water consumption, become more efficient and be drought-proof the Water Corporation has initiated its Water Efficiency Management Plan Program. The Water Efficiency Management Plan Program requires all businesses using more than 20,000 kL of water a year to complete a Water Efficiency Management Plan (WEMP) to help save water. The programme involves businesses detailing water saving actions and initiatives and providing annual progress reports about their efforts. As part of the programme a WEMP includes:

- Site water use history
- Water saving opportunities (including benchmark indicators and targets)

- Water saving action plan (including timeframes)
- Management and Water Corporation commitment

Once the WEMP is submitted and accepted the plan is valid for 5 years. However, if the business changes ownership or water use increases significantly a revised WEMP may need to be submitted (Water Corporation, 2016).

Water pricing

In urban water resource management, economic theory suggests that demand for water should act like any other goods: as price increases, water use decreases. Water managers can use a variety of different price structures, all of which send different conservation signals to individuals and communities. A flat rate is essentially a fixed charge for water usage regardless of the volume used, where typically the size of the charge is related to the customer's property value (Sibly, 2006; Policy Research Initiative, 2005). However, while fixed prices enable water utilities to raise sufficient revenue for the operation and maintenance of the water supply network it does not provide any incentive for individuals and households to conserve water (CAP-Net, 2008; Olmstead and Stavins, 2007). A volumetric rate is a charge based on the volume used at a constant rate. An increasing block tariff rate contains different prices for two or more pre-specified quantities (blocks) of water, with the price increasing with each successive block. A two-part tariff system involves a fixed and a variable component. In the fixed component, water users pay one amount independently of consumption to cover infrastructural and administrative costs of supplying water. Meanwhile, the variable amount is based on the quantity of water consumed and covers the costs of providing water as well as encouraging conservation.

Case study 2: Cape Town's water price

Because Cape Town is situated in a water-scarce region, the city imposes water restrictions on a permanent basis, with the level of water restrictions dependent on dam storage levels. Cape Town has three levels of water restrictions:

- Level 1 (10% water savings): Normally in place.
- Level 2 (20% water savings): Applicable when dam levels are lower than the norm.
- Level 3 (30% water savings): Applicable when dam levels are critically low.

Since 1 November 2016, Cape Town has been under Level 3 (30% water savings) which restricts water usage activities including the prohibition of residents using sprinkler systems, watering their gardens, and washing their cars with hosepipes with municipality-supplied drinking water. To reduce water consumption further, Cape Town will, from 1 December 2016 until further notice, charge all residential, commercial and industrial water uses Level 3 (30% savings) tariffs (City of Cape Town, 2016b). For domestic water users, the first 6,000 litres remains free, but the next block rates will increase significantly (Table 3.1). Meanwhile, commercial and industrial waters, who are normally charged R18.77 including VAT per thousand

Table 3.1 Cape Town's residential water tariffs.

Water 2016/17 (domestic full) Steps	Unit*	Level 1 (10% reduction) Normal tariffs Rands (incl. VAT)	Level 2 (20% reduction) During level 2 restrictions Rands (incl. VAT)	Level 3 (30% reduction) During level 3 restrictions Rands (incl. VAT)
Step 1 (> 0 ≤ 6 kl)	/kl	R 0.00	R 0.00	R 0.00
Step 2 (> 6 ≤ 10.5 kl)	/kl	R 14.89	R 15.68	R 16.54
Step 3 (> 10.5 ≤ 20 kl)	/kl	R 17.41	R 20.02	R 23.54
Step 4 (> 20 ≤ 35 kl)	/kl	R 25.80	R 32.65	R 40.96
Step 5 (> 35 ≤ 50 kl)	/kl	R 31.86	R 48.93	R 66.41
Step 6 (> 50 kl)	/kl	R 42.03	R 93.39	R 200.16

* *1 kl is a thousand litres* (City of Cape Town, 2016b)

Table 3.2 Cape Town's commercial and industrial water tariffs.

Commercial & Industrial water use (Standard)	Unit	Level 1 (10% reduction) Rands (incl. VAT)	Level 2 (20% reduction) Rands (incl. VAT)	Level 3 (30% reduction) Rands (incl. VAT)
Water	Per kl	R 18.77	R 21.82	R 25.35

(City of Cape Town, 2016a)

litres (under Level 1 water savings), will now be paying the level 3 charge of R 25.35 per thousand litres (Table 3.2) (City of Cape Town, 2016a).

Subsidies and rebates

Economic instruments such as subsidies and grants (incentives) or rebates are used to modify an individual's behaviour in a predictable, cost-effective way, i.e., reducing wastage and lowering water consumption by providing, for example, subsidies for newer, more water-efficient toilets (Global Water Partnership, 2012; Policy Research Initiative, 2005; Savenije and Van Der Zaag, 2002; Marchal et al., 2012). In particular, incentives are commonly used to encourage the uptake of water-efficient appliances, as positive incentives are found to reduce the gap between the time of the incentive is presented and behavioural change as compared to disincentives (Policy Research Initiative, 2005).

Case study 3: Los Angeles' water-energy grant

With climate change droughts and extreme weather events projected to lower water availability for cooling of thermoelectric power stations and impact hydropower generation, which in turn lowers the availability of electricity to import water, the Los Angeles Department of Water and Power (LADWP) has developed the Water Conservation Technical Assistance Program that offers commercial, industrial, institutional and multi-family customers a customised approach to reducing their water, and energy bills (LADWP, 2016). Through the programme LADWP works one-on-one with customers to modernise their facility with the latest water-efficient equipment. The programme offers up to US $250,000 in financial incentives for pre-approved equipment and products that demonstrate water savings. The actual incentive amount is based on the water savings accomplished by the project, with the incentive calculated at US $1.75 per 1,000 gallons of water saved over a number of years—not to exceed the installed cost of the project—with customers receiving a rebate following verification of installation and operation of pre-approved projects (LADWP, 2016).

Product labelling

Urban water managers can promote water conservation product labelling schemes as the labelling of household appliances according to their degree of water efficiency is important in reducing household water consumption by eliminating unsustainable products from the market, provided the labelling scheme is clear and comprehensible and identifies both the private and public benefits of conserving water. Nevertheless, people are more likely to respond to eco-labels if the environmental benefits closely match personal benefits, such as reduced water bills (Brears, 2014).

Case study 4: Australia's Water Efficiency Labelling Scheme

Australia's Water Efficiency Labelling Scheme (WELS) requires certain products to be registered and labelled with their water efficiency in order to save water, and energy. Under the scheme, over a third of water savings will come from more efficient showers, around 34% will come from washing machines and 23% from toilets and urinals. By 2021, it is projected that Australia could save more than one billion dollars through reduced water and energy bills by choosing more efficient products. It is also estimated that by 2021, water efficient products will help: reduce domestic water usage by more than 100,000 megalitres per annum; save more than 800,000 megalitres; and reduce total greenhouse gas output by 400,000 tonnes each year, equivalent to taking 90,000 cars off the road each year (Australian Goverment, 2016).

Retrofit programmes

Retrofit programmes involve the distribution and installation of replacement devices to physically reduce water use in homes and offices. The most common retrofits are toilet retrofits, involving customers having their older toilets replaced with newer low-/dual-flush toilets, and the distributing of showerheads and faucet aerators (devices that when inserted into taps reduce the flow of water) to households and offices (Georgia Environmental Protection Division Watershed Protection Branch, 2007; Michelsen et al., 1999; Pennsylvania State University, 2010). Water-saving devices can be distributed by water managers in numerous ways including door-to-door delivery of water-saving kits to households or direct installation by trained technicians or plumbers (Pennsylvania State University, 2010).

Case study 5: Anglian Water's free 'bits & bobs'

Anglian Water supplies water to the fastest growing region in the United Kingdom, which in some parts has less rainfall per year than Jerusalem. To reduce water scarcity, the utility is aiming to reduce customers' per capita consumption by 20 litres/day. One of the key programmes for reducing demand is Anglian Water's 'Bits and Bobs' water savings programme, which involves customers receiving a free water-saving home visit and the installation of free water-saving devices. The programme involves the utility visiting over 20,000 homes across the service region, with average customers expected to save over 40 litres of water per household per day (Anglian Water, 2016a). This translates into an average saving on their water bill of £40 in addition to further savings on household energy bills. To ensure the water saving devices are properly installed a qualified member of Anglian Water will visit the homes and install the devices as well as provide tips on how to save water (Anglian Water, 2016b).

Communication and Information Instruments

Communication and information instruments encourage a water-orientated society. In particular, communication and information tools aim to change behaviour through public awareness campaigns around the need to conserve scarce water resources as well as school education programs that teach young people on the need to conserve water.

Public education

Urban water managers can use public education to persuade individuals and communities to conserve water resources. In particular, water managers can influence individual's attitudes and behaviours towards water resources by increasing their knowledge and awareness of environmental problems associated with water scarcity (Steg and Vlek, 2009; Najjar and Collier, 2011; Policy Research Initiative, 2005). Meanwhile, urban water managers can promote water conservation in schools to increase young people's knowledge of the water cycle and encourage the sustainable use of scarce resources (Brears, 2016). To do so, water managers can use a variety of strategies, including school presentations, and distribution of water conservation information and materials that can be used in school curriculum. There are multiple tools and formats that water managers can use to increase environmental awareness and water conservation that are summarised in Table 3.3.

Case study 6: Irish Water supporting the Green-Schools programe

Over the past four years, Irish Water has sponsored and supported the country's Green-Schools programme which aims to develop awareness around water conservation in both schools and homes. As part of Green-Schools, Irish Water runs the ambassador programme in which the utility directly engages with second-level students on the topics of water, water conservation, treatment and the marine environment and

Table 3.3 Demand management educational tools.

Tool	Description
Public information	Printed literature distributed or available for the public, public service announcements and billboard advertisements, public transportation, television commercials, newspaper articles and advertisements and Internet and social media campaigns
Public events	Conservation workshops and public exhibitions where customers can receive information on both water conservation tips and the various types of water-saving devices available
Information in water utility bills	Water bills should be understandable, enabling customers to easily identify the usage volume, rates and charges. The water bill should be informative, enabling customers to compare their current bill with previous bills. The water bill can also contain leaflets on water conservation tips

(U.S. EPA, 1998; Pennsylvania State University, 2010; The State of Israel Ministry of National Infrastructures Planning Department Water Authority, 2011; Georgia Environmental Protection Division Watershed Protection Branch, 2007; Keramitsoglou and Tsagarakis, 2011)

encourages them to act as ambassadors for Green-Schools in their own schools and local communities. In addition, Irish Water supports the Green-Schools Water of the Year Award, which recognises innovation and creativity among schools in spreading awareness about water issues (Green-Schools, 2016).

Competitions between water users

Urban water managers can increase participation rates in water conservation programmes by promoting competition among individuals and communities to achieve specific water consumption targets. Examples of competitions include eliciting commitments to water savings targets and promoting competition through the water utility bill. Regarding elicitation of commitments, urban water managers can obtain verbal or written commitments from individuals and communities in order to achieve specific water saving targets. Competitions can then be held to compare individuals' or communities' water savings with one another and offer winners recognition or prizes for their water saving achievements. The water bill can also be used as a tool for competition between water users; for example, water bills can show a household's water consumption compared to the average household in the neighbourhood, city, province or state (Georgia Environmental Protection Division Watershed Protection Branch, 2007; Patchen, 2010).

Case study 7: New York City's Water challenge to hotels

Over the period 2013–2014, New York City launched its Water challenge to hotels in which some of the city's largest hotels volunteered to participate in the challenge. The Water challenge involved participating hotels reducing their water usage per square foot by 5% in one year. The baseline for comparison was the average daily water use per square foot, which was developed for each hotel utilising all available water consumption data during a 12-month period prior to the start of the challenge. The water challenge required that participants: be responsible for tracking and monitoring water use; develop a water conservation plan that mapped out a strategy to reduce water usage; and implement the water conservation and water efficiency projects outline in their water conservation plans. The water challenge was successful with four hotels managing to surpass the 5% goal and save more than 10% over the previous one year period (New York DEP, 2016).

Corporate social responsibility

Water utilities can show leadership in conservation for several reasons. First, a failure to exemplify the behavioural changes water utilities wish to see in their customers will undermine any information campaigns water utilities are attempting to engage in at a future date, second, successful internal conservation programmes send a strong signal to water customers about what is possible and that water utilities are serious about water conservation and third, these initiatives allow water utilities to learn invaluable lessons first-hand on the difficulties of achieving water conservation goals (Jackson, 2005).

Case study 8: Sydney Water leading the way

Sydney Water faces multiple challenges that threaten its ability to continue delivering water at affordable prices including population growth, concerns about the cost of living, greater customer expectations and climate change. To meet these challenges the utility aims to become more efficient in providing high quality water and services to its customers. As part of this efficiency drive, Sydney Water is proposing to lower its prices for customers over the 2016–2020 period, resulting in most households saving over A$100 on their water and wastewater bills each year (Sydney Water, 2016a; Sydney Water, 2016b). To increase efficiency, reduce greenhouse gas emissions and pass on the savings to customers, Sydney Water is reducing its energy use by improving energy efficiency in its operations and generating renewable energy. The aim is to keep its non-renewable energy purchases at or below 1998 levels as well as cap its carbon emissions at a stable level. To date the utility has saved almost 30 GWh of energy: the equivalent of saving electricity used by 4,100 homes a year. This has included: Using smarter mixing techniques at its wastewater treatment plants (WWTPs); minimising power use by aerators at WWTPs; investing in energy efficient buildings; and replacing conventional lighting with LED technology at several sites, including WWTPs, saving around A$130,000 a year (Sydney Water, 2016a).

Conclusions

With global demand for water projected to outstrip supply in the coming decades as a result of rapid urbanisation, rising water-energy nexus pressures and climate change, urban centres around the world need to transition towards a demand management framework that balances rising demand for water with often limited, and variable, supplies.

In a transition towards managing natural resources sustainably, resource managers are faced with climatic and non-climatic drivers. Regarding climate change, natural resources managers can transition from hard solutions that mitigate the impacts of climate change towards soft actions that focus on altering human behaviour as a way of reducing vulnerability to climatic risks. Faced with resource scarcity due to various non-climatic trends, natural resources managers can transition from the typical reliance on supply-side projects towards strategies that focus on behavioural change as a way of decreasing demand for scarce resources.

In the context of water resources management, cities facing rising demand for limited, and often variable, supplies of water can transition away from costly traditional supply-side infrastructural projects towards managing actual demand for water, where demand management involves better use of existing water supplies before plans are made to further increase supply. With demand management involving communicating ideas, norms and innovations for water conservation, urban water managers can use a variety of regulatory and economic instruments as well as communication and information instruments to encourage the conservation and efficient use of water to reduce urban water-energy-climate nexus pressures.

Regarding regulatory and economic instruments, utilities and planners around the world have initiated a variety of demand management tools. For instance, conservation ordinances and restrictions can be used to restrict water usage temporarily or permanently, with one example being Western Australia's Water Corporation mandating water efficiency management plans for large water users to encourage businesses to become drought-proof while at the same time lowering their operational costs. Regarding the pricing of water, basic economic theory suggests that as price increases, water consumption decreases with various pricing structures available to utilities to encourage conservation. For example, Cape Town has developed a pricing system with water tariffs that increase with rising levels of water scarcity. Meanwhile, utilities use subsidies and rebates to modify behaviour in a cost-effective way: LADWP has developed customised incentives for large-water users to reduce their water and energy bills by modernising their facilities with the latest water-efficient equipment. To encourage the purchasing of water-efficient products that also save energy, many jurisdictions have implemented product labelling schemes, for example Australia's WELS scheme requires certain products to be registered and labelled to reduce water and energy consumption in addition to reducing greenhouse gas emissions. To encourage water and energy savings at homes, water managers have implemented retrofit programmes that give away water-saving devices for homes. For instance, Anglian Water's retrofit programme involves customers receiving a free water saving home visit and the installation of free water saving devices.

Urban water managers can also use communication and information instruments to change the behaviour of water users. For instance, public education via informative publications, public events and leaflets in water bills can inform customers on how to conserve water at home as well as receive information on various water-saving devices on the market. Water managers can also provide educational materials and visit classrooms to educate young people on the need to conserve water and related scarce resources. Irish Water is one leading example of a utility encouraging young people to save water and reduce environmental degradation, with the utility engaging students in the classroom on water and related topics as well as sponsoring a Water of the Year Award that recognises innovation and creativity among schools in spreading awareness on water-related issues. To encourage participation in water conservation programmes, water managers can also initiate competitions among individuals and communities to achieve specific water consumption targets. For example, New York City initiated a water challenge to hotels to encourage lower water consumption levels in this sector. Finally, urban water managers can show leadership in conservation to exemplify the behaviour change it aims to achieve in water users, sending a strong signal to water customers that the utility is serious about conservation and learning first-hand the challenges of achieving conservation goals. For example, Sydney Water is aiming to reduce its energy costs in providing water services as well as lower its greenhouse gas emissions, with the associated efficiency gains passed onto customers in the form of lower water bills.

Overall, cities are likely to face water scarcity from urban population growth, rising water-energy nexus pressures and climate change in the coming decades. Therefore, urban centres will need to transition towards a demand management framework that involves the use of regulatory and economic instruments, including conservation ordinances, pricing of water, subsidies and rebates, product labelling and retrofit programmes, as well as communication and information instruments including public education, school education and leading-by-example to modify the attitudes and behaviour of water users to balance rising demand for limited, and often variable, supplies of water.

References

ADB. 2013. 3 in 4 Asia-Pacific Nations Facing Water Security Threat—Study [Online]. Available: https://www.adb.org/news/3-4-asia-pacific-nations-facing-water-security-threat-study.

Adger, W.N., Agrawala, S., Mirza, M.M.Q., Conde, C., O'brien, K., Pulhin, J., Pulwarty, R., Smit, B. and Takahashi, K. 2007. Assessment of adaptation practices, options, constraints and capacity. *In*: Parry, M.L., Canziani, O.F., Palutikof, J.P., Van Der Linden, P.J. and Hanson, C.E. (eds.). Climate Change 2007: Impacts, Adaptation and Vulnerability. Cambridge: Cambridge University Press.

Anglian Water. 2016a. Frequently asked questions [Online]. Available: http://www.anglianwater. co.uk/environment/how-you-can-help/using-water-wisely/we-products/faqs.aspx.

Anglian Water. 2016b. Save bucket loads with a free water saving home visit [Online]. Available: http://www.anglianwater.co.uk/environment/how-you-can-help/using-water-wisely/we-products/.

Australian Goverment. 2016. About WELS [Online]. Available: http://www.waterrating.gov.au/about-wels.

Australian Government Productivity Commission. 2012. Barriers to effective climate change adaptation. Available: http://www.pc.gov.au/inquiries/completed/climate-change-adaptation/report/climate-change-adaptation.pdf.

Bithas, K. 2008. The sustainable residential water use: sustainability, efficiency and social equity. The European experience. Ecological Economics, 68: 221–229.

Bahri, A. 2012. Integrated urban water management. https://www.gwp.org/globalassets/global/toolbox/publications/background-papers/16-integrated-urban-water-management-2012.pdf.

Brears, R. 2014. Transitioning Towards Urban Water Security in Asia Pacific. Available: http://www.diss.fu-berlin.de/docs/receive/FUDOCS_document_000000021440?lang=de.

Brears, R.C. 2016. Urban Water Security, Chichester, UK; Hoboken, NJ, John Wiley & Sons.

Canada West Foundation. 2004. Drop by drop: Urban water conservation practices in Western Canada. Available: http://cwf.ca/research/publications/drop-by-drop-urban-water-conservation-practices-in-western-canada/.

CAP-NET. 2008. Economics in sustainable water management. Available: http://www.cap-net.org/training-material/economics-in-sustainable-water-management-english/.

City of Cape Town. 2016a. Commercial water restrictions explained [Online]. Available: http://www.capetown.gov.za/Work%20and%20business/Commercial-utility-services/Commercial-water-and-sanitation-services/2016-commercial-water-restrictions-explained.

City of Cape Town. 2016b. Residential water restrictions explained [Online]. Available: http://www.capetown.gov.za/Family%20and%20home/residential-utility-services/residential-water-and-sanitation-services/2016-residential-water-restrictions-explained.

European Environment Agency. 2013. Adaptation in Europe: Addressing risks and opportunities from climate change in the context of socio-economic developments. Available: http://www.eea.europa.eu/publications/adaptation-in-europe.

FAO. 2016. Coping with water scarcity in agriculture: A global framework for action in a changing climate Available: http://www.fao.org/3/a-i5604e.pdf.

Georgia Environmental Protection Division Watershed Protection Branch. 2007. Water conservation education programs. Available: http://www1.gadnr.org/cws/Documents/Conservation_Education.pdf.

Gifford, R., Kormos, C. and Mcintyre, A. 2011. Behavioral dimensions of climate change: drivers, responses, barriers, and interventions. *Wiley Interdisciplinary Reviews: Climate Change* 2: 801–827.

Global Water Partnership. 2012. Water demand management: The Mediterranean experience. Available: http://www.gwp.org/Global/ToolBox/Publications/Technical%20Focus%20Papers/01%20Water%20Demand%20Management%20-%20The%20Mediterranean%20Experience%20(2012)%20English.pdf.

Green-Schools. 2016. Green-Schools and Irish Water launch fourth year of partnership [Online]. Available: https://greenschoolsireland.org/green-schools-and-irish-water-launch-fourth-year-of-partnership/.

Hoffman, A.J. 2010. Climate change as a cultural and behavioral issue: Addressing barriers and implementing solutions. *Organizational Dynamics* 39: 295–305.

ICPDR. 2016. Climate change adaptation [Online]. Available: https://www.icpdr.org/main/activities-projects/climate-change-adaptation.

IEA. 2016. A delicate balance between water demand and the low-carbon energy transition [Online]. Available: http://www.iea.org/newsroom/news/2016/november/a-delicate-balance-between-water-demand-and-the-low-carbon-energy-transition.html.

Jackson, T. 2005. Motivating sustainable consumption. *Sustainable Development Research Network* 29: 30.

Keramitsoglou, K.M. and Tsagarakis, K.P. 2011. Raising effective awareness for domestic water saving: evidence from an environmental educational programme in Greece. *Water Policy* 13: 828–844.

Kolikow, S., Kragt, M.E. and Mugera, A.W. 2012. An interdisciplinary framework of limits and barriers to climate change adaptation in agriculture. Available: https://ideas.repec.org/p/ags/uwauwp/120467.html.

LADWP. 2016. Custom water conservation projects (TAP) [Online]. Available: https://www.ladwp.com/ladwp/faces/wcnav_externalId/a-w-cstm-wtr-prjct-tap?_adf.ctrl-state=kg8odqt9w_30&_afrLoop=526711373240641.

Lieberherr-Gardiol, F. 2008. Urban sustainability and governance: issues for the twenty-first century. *International Social Science Journal* 59: 331–342.

Maheepala, S., Blackmore, J., Diaper, C., Moglia, M., Sharma, A. and Kenway, S. 2010. Towards the adoption of integrated urban water management approach for planning. *Proceedings of the Water Environment Federation,* 6734–6753.

Marchal, V., Dellink, R., Van Vuuren, D., Clapp, C., Château, J., Lanzi, E., Magné, B. and Van Vliet, J. 2012. OECD environmental outlook to 2050: the consequences of inaction. Paris: OECD Publishing.

Michelsen, A.M., Mcguckin, J.T. and Stumpf, D. 1999. Nonprice water conservation programs as a demand management tool. *JAWRA Journal of the American Water Resources Association* 35: 593–602.

Milbrath, L.W. 1995. Psychological, cultural, and informational barriers to sustainability. *Journal of Social Issues* 51: 101–120.

Molle, F. and Berkoff, J. 2009. Cities vs. agriculture: A review of intersectoral water re-allocation. Natural Resources Forum. Wiley Online Library, 6–18.

Muller, M. 2007. Adapting to climate change water management for urban resilience. *Environment and Urbanization* 19: 99–113.

Najjar, K. and Collier, C.R. 2011. Integrated water resources management: bringing it all together. *Water Resources Impact* 13: 3–8.

New York Dep. 2016. New York City water challenge to hotels [Online]. Available: http://www.nyc. gov/html/dep/html/ways_to_save_water/nyc-water-challenge.shtml.

OECD. 2012. OECD Environmental outlook to 2050: The consequences of inaction highlights. Available: https://www.oecd.org/env/indicators-modelling-outlooks/49846090.pdf.

Olmstead, S.M. and Stavins, R.N. 2007. Managing water demand: price vs. non-price conservation programs. *A Pioneer Institute White Paper,* 39.

Patchen, M. 2010. What shapes public reactions to climate change? Overview of research and policy implications. *Analyses of Social Issues and Public Policy* 10: 47–68.

Pennsylvania State University. 2010. Water conservation for communities. Available: http:// extension.psu.edu/natural-resources/water/conservation/water-conservation-home-study/ why-conserve-water/communitywaterconservation.pdf.

Policy Research Initiative. 2005. Economic instruments for water demand management in an integrated water resources management framework. Ottawa, ON [Online]. Available: http:// publications.gc.ca/collections/Collection/PH4-18-2005E.pdf.

Savenije, H.H. and Van Der Zaag, P. 2002. Water as an economic good and demand management paradigms with pitfalls. *Water International* 27: 98–104.

Sibly, H. 2006. Efficient urban water pricing. *Australian Economic Review* 39: 227–237.

Steg, L. and Vlek, C. 2009. Encouraging pro-environmental behaviour: An integrative review and research agenda. *Journal of Environmental Psychology* 29: 309–317.

Sydney Water. 2016a. Energy management and climate change [Online]. Available: http://www. sydneywater.com.au/SW/water-the-environment/what-we-re-doing/energy-management/ index.htm.

Sydney Water. 2016b. Summary annual report 2015–2016. Available: http://www.sydneywater.com. au/web/groups/publicwebcontent/documents/document/zgrf/mdk1/~edisp/dd_095615.pdf.

The State of Israel Ministry of National Infrastructures Planning Department Water Authority. 2011. The State of Israel: National water efficiency report. Available: http://www.water.gov. il/Hebrew/ProfessionalInfoAndData/2012/24-The-State-of-Israel-National-Water-Efficiency-Report.pdf.

Times of India. 2013. 22 of India's 32 big cities face water crisis [Online]. Available: http://timesofindia.indiatimes.com/india/22-of-Indias-32-big-cities-face-water-crisis/ articleshow/22426076.cmss.

U.S. EPA. 1998. Water conservation plan guidelines. Available: https://www3.epa.gov/watersense/ pubs/guide.html.

UN-Water. 2014. Water, energy and food are inextricably linked [Online]. Available: http://www. unwater.org/statistics/statistics-detail/en/c/246966/.

UN. 2014. World's population increasingly urban with more than half living in urban areas [Online]. Available: http://www.un.org/en/development/desa/news/population/world-urbanization-prospects-2014.html.

UNEP. 2016. Half the world to face severe water stress by 2030 unless water use is "decoupled" from economic growth, says International Resource Panel [Online]. Available: http://www. unep.org/NewsCentre/default.aspx?DocumentID=27068&ArticleID=36102.

United Nations. 2014. World's population increasingly urban with more than half living in urban areas [Online]. Available: http://www.un.org/en/development/desa/news/population/world-urbanization-prospects-2014.html.

United Nations Population Fund. 2016. Urbanization [Online]. Available: http://www.unfpa.org/urbanization.

Van Roon, M. 2007. Water localisation and reclamation: steps towards low impact urban design and development. *Journal of Environmental Management*, 83: 437–447.

Water Corporation. 2016. Water efficiency management plan programe [Online]. Available: https://www.watercorporation.com.au/home/business/saving-water/water-efficiency-programs/water-efficiency-management-plan.

Williams, C.C. and Millington, A.C. 2004. The diverse and contested meanings of sustainable development. *The Geographical Journal* 170: 99–104.

Wolfe, S. and Brooks, D.B. 2003. Water scarcity: An alternative view and its implications for policy and capacity building. Natural Resources Forum, Wiley Online Library, 99–107.

World Bank. 2007. Making the most of scarcity: Accountability for better water management results in the Middle East and North Africa. Available: http://siteresources.worldbank.org/INTMNAREGTOPWATRES/Resources/Making_the_Most_of_Scarcity.pdf.

World Bank. 2014. Africa's urban population growth: trends and projections. Available: http://blogs.worldbank.org/opendata/africa-s-urban-population-growth-trends-and-projections.

Section 2
Case Study

Chapter-4

The Water-Energy-Food Nexus in Africa

Genesis T. Yengoh

INTRODUCTION

The Bonn 2011 Nexus Conference, "The Water Energy and Food Security Nexus—Solutions for the Green Economy" laid the groundwork for recognizing the complex interactions between the components of Water-Energy-Food (WEF) nexus and global goals, as well as challenges of achieving sustainable development (Hoff, 2011). Water, energy and food security are essential needs for human societies worldwide (Karabulut et al., 2015), and central to the functioning of modern life. These three elements are fundamental to the social and economic development of countries and regions and can be a major determinant of political stability in others. The need to minimize trade-offs and risks of adverse cross-sectoral impacts when approaching WEF components is essential in sustaining the natural systems that support the production of these products and services (Hoff, 2011).

The water, energy, and food systems are interconnected. The nexus is the link or relationship which explains how and where these three systems intersect. As a result of the close interconnectedness of these systems, policies, and actions related to one system can impact one or both of the other systems. The nexus approach offers the opportunity to develop a strong understanding of the interdependencies among these three systems, and to guide integrated planning, policy, and management in these sectors.

Bazilian et al. (2011) found close associations between elements in the WEF nexus. Among other similarities, all three areas are experiencing resource constraints and rapidly growing global demand. They are all "global goods"—involved in international trade with global implications, have different regional availability, have strong interdependencies with climate change, and raise deep security issues as they

Lund University Centre for Sustainability Studies – LUCSUS, Biskopsgatan 5, SE – 223 62 Lund, Sweden.
Email: yengoh.genesis@lucsus.lu.se

are all fundamental to the functioning of society (Bazilian et al., 2011). Given the close bonds between water, energy, and food, the policy areas that address these key sectors of human well-being have been described to have numerous interwoven concerns (Bazilian et al., 2011). These include access to products and services, environmental impacts, price volatility, issues of equity, security concerns, and implications for sustainable development (Bazilian et al., 2011).

To better address the conflicts that occur as these different spheres of human development interact, there is a need for the development of new institutional capacity, as well as the enhancement of viable existing structures of cooperation and collaboration to better harness the outcomes of the WEF nexus. Transformative social and political change is also needed to create new structures, markets, and governance to deal with the WEF nexus in order to meet a range of sustainable development goals (Keairns et al., 2016).

In the past, sectoral approaches have dominated the management of water, energy, and food (Smajgl et al., 2016). Such approaches led to fragmented and uncoordinated development and management of these resources, limiting the possibilities for a holistic appreciation of the complex relationships and interactions that exist between them. A nexus approach permits a better understanding of these complexities and provides tools for a more sustainable management of our water, energy and food resources (Howarth and Monasterolo, 2016). A nexus approach enables us to better understand the competition, anticipate potential trade-offs and synergies, resolve conflicts, as well as appraise and prioritize response options that may arise in these critical resource sectors.

Water, Energy, and Food in Africa

Water resources—demand, use and future challenges

Agriculture is the biggest consumer of the world's freshwater, accounting for almost 70% of all withdrawals, and up to 95% in developing countries. The FAO estimates that over the next 30 years, 70% of gains in cereal production will come from irrigated land, meaning that the demand for global freshwater resources will become even greater (Alexandratos and Bruinsma, 2012). One of the biggest challenges for water managers both at present and into the future is, therefore, to increase the efficiency of irrigated agriculture and reduce waste (Bach et al., 2012). This will provide possibilities for increasing food production with minimal effects on water needs for other sectors of the economy and the environment. The world could celebrate that the MDG target for safe drinking water was met in 2010, well ahead of the deadline of 2015, with over 90% of the world's population having access to improved sources of drinking water (UNICEF and WHO, 2015). However, progress was not the same with regards to the sanitation component of water, as 1.8 billion people continue to rely on a source of drinking water that is contaminated with fecal matter (UNICEF and WHO, 2015).

Water is one of the key resources required to sustain the health and support the functioning of ecosystems. Ecosystems tend to be highly sensitive to the alteration of hydrological flows and the quality of water in them. Alterations of hydrological flows can result from excessive withdrawal of water from rivers, lakes, and aquifers, deforestation of catchments or the drainage of riparian buffers which hold and sustain water resources, as well as from changes in the catchment land cover. The quality of water in the hydro-system may be affected by the discharge of poorly treated wastewater or saline brine into natural bodies, runoff from human activities such as agriculture and mining, and the build-up of contaminants resulting from reductions in volume. Natural cases of contamination may occur from the leaching of contaminants such as arsenic into local water bodies.

Access to freshwater in the right quantity and quality and food security are therefore closely connected. It has been observed that people with better access to water tend to have lower levels of undernourishment (FAO, 2008b). Since water is a key determinant of food security, lack of it can be one of the leading causes of famine and undernourishment, especially in areas where people depend on rain-fed agriculture for food and livelihoods. The depletion and seasonal scarcity of water is a common feature in the Sudano-Sahelian belt of Africa, and more recently in many regions of southern Africa (Fig. 4.1). In sub-Saharan Africa, manifestations of extreme weather events associated with water such as erratic rainfall and seasonal differences in water availability, floods and droughts have been known to cause some of the most intensive food emergencies (FAO, 2008b). Also, in developing regions of the world, such as sub-Saharan Africa, drought ranks as the single most common cause of severe food shortages (FAO, 2008b). Irrigation has the potential of increasing yields of most food crops by 100 to 400%. In sub-Saharan Africa, only about 5% of the land is irrigated even though about 39% of the land area is suitable for irrigation (FAO, 2008b). Increasing the productivity of agriculture through better

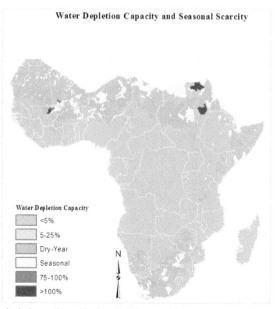

Figure 4.1 Water depletion and scarcity in sub-Saharan Africa (Data source: Brauman et al. (2016)).

Color version at the end of the book

water management can make a significant impact on agricultural productivity and the livelihoods of farmers (Faurès and Santini, 2008).

Energy resources—demand, use and future challenges

The United Nations (2010) defines energy security as "access to clean, reliable and affordable energy services for cooking and heating, lighting, communications and productive uses." Clean, efficient, affordable and dependable energy is one of the most critical resources that provide the groundwork for and sustains the social, economic, environmental integrity of countries and regions. To increase productivity in different economic and social sectors, enhance competitiveness, reduce poverty and achieve economic growth, developing countries in particular need to expand access to reliable, modern and sustainable energy resources and services. In sub-Saharan Africa (excluding South Africa) just about 24% of the population has access to electricity, compared to 40% in other low-income countries. Manufacturing enterprises experience power outages on average 56 days per year, and energy tariffs are generally high with an average of US$0.13 per kilowatt-hour, compared to most parts of the developing world where tariffs fall in the range of US$0.04 to US$0.08 per kilowatt-hour (World Bank, 2016). In the west of the continent, the diffusion of electricity transmission networks is limited to mainly coastal regions which harbor a majority of the economic activities of most countries open to the Atlantic Ocean (Fig. 4.2a). While the number of power stations serving the continent remains relatively limited, there is enormous potential for the development of small and mini hydropower plants that have the potential of contributing significantly in mitigating

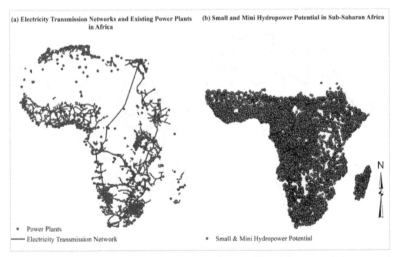

Figure 4.2 The energy situation and potentials of Africa: (a) Electricity transmission networks and existing power plants in Africa; and (b) Potential for small and **mini-hydro power** production in Sub-Saharan Africa (Data sources: the Africa Electricity Transmission Network (AICD 2009) and studies carried out by Dimitris Mentis of the Division of Energy Systems Analysis, Royal Institute of Technology in Stockholm, Sweden).

Color version at the end of the book

the problem of access to electrical energy on the continent (Fig. 4.2b). Currently, an estimated 1.2 billion people (about 17% of the global population) lack electricity, and 2.7 billion people (about 38% of the world's population) continue to put their health at risk through reliance on the traditional use of solid biomass for cooking (IEA, 2015). Such biomass fuels include fuelwood, charcoal, agricultural waste and animal dung, to meet their energy needs for cooking. Access to and the use of energy directly affects the ability to produce and distribute potable water, as well as for the mechanical energy required for different stages of the production of agricultural goods and services.

The World Bank estimates that approximately 1.1 billion people are living without access to electricity, while another 2.9 billion use wood or other forms of biomass for cooking and heating. These forms of biomass fuels contribute to air pollution and is associated with health risks—evidenced by contributing to the deaths of some 4.3 million people annually (World Bank, 2015).

Food production, demand, and challenges

Having access to sufficient nutritious food is set out in the UN declaration of human rights. By 2050, the world's population is expected to grow by 2 billion people, up from 7.4 billion in 2015 (UN-DESA, 2015). Besides the increase in population numbers which entail more mouths to feed, there is an ongoing trend of changes in dietary patterns with an increasing share of animal products in a growing proportion of the world's population. The FAO estimates that the world's growing population will require about 50% more food by 2030 compared to 1998. Feeding this growing population and reducing hunger can only be possible if agricultural yields can be increased significantly and sustainably. Unfortunately, the gaps between current and potential yields for major food crops remain significant in sub-Saharan Africa (Fig. 4.3).

In 2015, it was reported that globally, approximately 795 million people were undernourished (FAO, IFAD, and WFP, 2015). This represented a decline of 167 million over the last decade, and 216 million less than in 1990–92. While most of this reduction has occurred in developing countries, in sub-Saharan Africa, progress has been slow overall, despite many success stories at national and sub-regional levels. In some cases, the modest gains made by a number of countries have been wiped out by factors such as civil and political unrests which have in some cases even contributed to raising the ranks of the hungry (FAO et al., 2015). Investment in agriculture, evidenced by the contribution of agricultural expenditure as a share of the Gross Domestic Product saw a decline following the 2008–09 global financial crisis and is still to recover in many regions of the continent (Fig. 4.4). As the world adopted the Sustainable Development Goals (SDGs) in September of 2015, it was noted that progress on agriculture-related MDGs had not been uniform across the African continent (FAO, 2015). The FAO (2015) reported that achievement of Goal 1 of eradicating extreme poverty and hunger had been modest at best. Three countries reduced hunger by 50% or more (Ghana, DRC & Mauritania); 19 countries reduced it by 20–49.9%; and 13 by 0.0–19.9%. Five countries (Burundi, Swaziland, Comoros,

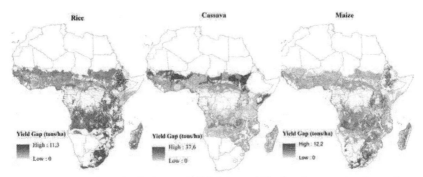

Figure 4.3 Yield gaps for major food crops in Sub-Saharan Africa based on growing degree days and precipitation (Data source: Foley et al. (2011) and Mueller et al. (2012).

Color version at the end of the book

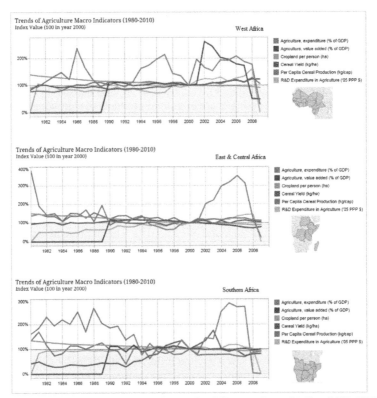

Figure 4.4 Trends of agricultural macro indicators for different regions of Africa (1980–2010) (Data source: FAOStat, ReSAKSS, ASTI, IFPRI, HarvestChoice).

Color version at the end of the book

Cote d'Ivoire and Botswana) experienced setbacks. Progress in World Food Security (WFS) targets was also not uniform. Countries that had achieved both MDG and WFS included Angola, Cameroon, Djibouti, Gabon, Ghana, Mali and Sao Tome and

Principe. Those that achieved MDG target and made progress towards WFS were Benin, Ethiopia, The Gambia, Malawi, Mauritania, Mauritius, Mozambique, Niger, Nigeria, Senegal, S. Africa and Togo. Those that made progress at least in one or both of the MDG and WFS included Botswana, Burkina Faso, Cape Verde, Chad, Congo, Guinea, Guinea-Bissau, Kenya, Lesotho, Rwanda, Sierra Leone and Zimbabwe.

The water, energy and food nexus—an African perspective

One of the highlights of the Bonn 2011 Nexus Conference, "The Water Energy and Food Security Nexus—Solutions for the Green Economy" was the call for a more integrated approach to understanding the interactions between food, water, and energy, as well as achieving security in these sectors as an important aspect of moving towards a green economy. The Bonn 2014 Conference, "Sustainability in the Water-Energy-Food Nexus," built on the 2011 meeting. It emphasized the importance for the coherence of cross-sector policy efforts and international cooperation for the successful governance of risks to the management and sustainable supply of water, energy, food and ecosystem services. Between these two events, several initiatives were undertaken by a diverse range of actors around the world to develop understanding, define policies, and set objectives within the context of the WEF nexus (Bizikova et al., 2013).

Understanding the African context in existing WEF nexus frameworks

The range of initiatives since the Bonn 2011 Nexus Conference has led to the development of a plethora of conceptual and analytical frameworks to guide the understanding, coordination, management and use of water, energy and food resources across sectors and scales. Some of the main frameworks have benefited from detailed reviews (Bizikova et al., 2013; ESCWA, 2015). While all of these frameworks can be applied to different aspects of the African setting, two of them (by Hoff et al., 2011; FAO, 2014) address some of the key issues and challenges unique to the African context.

The water, energy and food security nexus presented by Hoff (2011) identifies some of the key global trends that the African continent has to overcome in achieving security within WEF sectors: population growth, urbanization, and climate change (Fig. 4.5). Population growth is a key challenge for the continent. Africa's population is the fastest growing globally (UNO, 2015), with the continent alone contributing about 33% of the 82 million people added to the world's population annually (Adams and Opoku, 2016). Between 2015 and 2050, about 1.3 billion will be added in Africa (UNO, 2015).

This rapid population growth has been associated with urbanization which has been a key feature of Africa's demographic history since the 1960s (Fox, 2012). Africa is going to be the continent that suffers the most in terms of negative impacts of climate change on a range of environmental, economic, social and development sectors (UNFCCC, 2007). The framework (Fig. 4.5) also identifies the importance of governance, institutions and policy coherence in guiding investments and innovations in the WEF sectors to ensure that negative externalities across sectors are

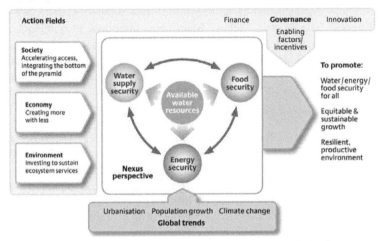

Figure 4.5 The water, energy and food security nexus (Hoff, 2011).

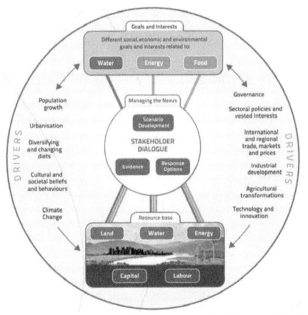

Figure 4.6 The FAO approach to the Water-Energy-Food Nexus (FAO, 2014).

minimized, benefits are equitably shared, and human rights are secured (Hoff, 2011). Governance within this context has been a recurring theme regarding challenges to the achievement of the sustainable development goals (UNECA, 2013b).

The WEF nexus framework proposed by FAO (2014) also suggests elements that respond to a number of features that characterize the African context. Besides population, urbanization and climate change, FAO (2014) also proposes cultural and societal beliefs, as well as diversifying and changing diets, and agricultural transformations as major drivers of dynamics in the WEF sectors (Fig. 4.6). Colding and Folke (2001) observed that taboos and societal norms are prevalent in many cultures, and serve substantially in guiding people's attitudes on how to interact with the natural environment.

These "resource and habitat taboos" (Colding and Folke, 2001) are prevalent in Africa and can determine the success or failure efforts towards managing and using WEF components. Africa's middle class is growing (Handley, 2015), and by so-doing, changing the landscape of consumption in the continent. In keeping with Bennett's Law, there will be a shift in the primary source of calories from starchy staples, to diverse diets that include more fat, meat, and fish as well as fruits and vegetables. This will put pressure on the water and energy resources required to produce such high energy, high-calorie products. The importance of labor as a resource base is also highlighted in the nexus (FAO, 2014). This is pertinent to the African context because labor is one of the resources that the continent currently has in plentiful supply, and will continue to do so for a long time to come.

A WEF nexus framework for Africa

The ultimate aim of achieving security in the WEF nexus components is to attain a sustainable level of human and environmental welfare. Key global indicators of this welfare (human development indices) remain low for much of Africa (UNDP). In understanding the African context of the WEF nexus, it is important to highlight the characteristics which make the situation of the region unique within the different nexus components (Fig. 4.7). The opportunities offered by the region are unique— the natural resource base, abundant labor, as well as the buoyant demographic growth accompanied by a growing middle class of consumers. Another opportunity is that there is potential for the region to leapfrog into cleaner technologies in its development future. Distinctive factors that characterize the WEF nexus operating environment in Africa include cultural norms, low levels of economic endowment of

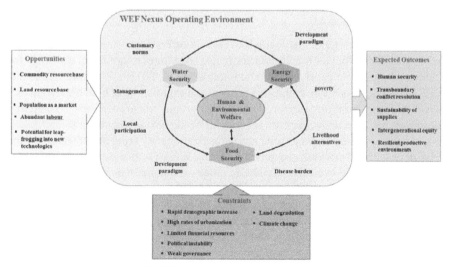

Figure 4.7 WEF nexus in the context of sub-Saharan Africa.

its populations, development paradigm of its countries, levels of local participation in the management of natural resources, markets, equity (gender, urban/rural, age-specific) in access to resources. Rapid demographic growth, rapid urbanization, low levels of infrastructure development, political instability, weak governance, limited capital, climate change, and land degradation are among the key constraints to the achievement of security in the WEF nexus components on the continent. Broader outcomes from the attainment of the WEF nexus component securities for Africa would include increased levels of human welfare, sustainability of supplies, transboundary conflict resolution, intergenerational equity, resilient, productive environments, and increased adaptive capacity to global environmental challenges (Fig. 4.7).

Opportunities: Africa generally has a rich natural resource base. It has 12% of the world's oil reserves, 40% of its gold, and 80 to 90% of its chromium and platinum (UNECA, 2013a). The continent also has about 60% of underutilized arable lands where yield gaps for major food crops remain significant (Yengoh and Ardö, 2014). It has been observed that this vast natural resource base can be harnessed for the development of the continent. Policies that promote commodity-based social and economic growth can contribute to the design and implementation of development in major sectors and nexuses such as water, energy, and food. Besides the natural resources, the continent's buoyant demographic growth has endowed it with a large labor force that can be harnessed for development. Besides providing labor for development, this population can also provide a market for resources and services on which the development of water food and energy can be based. It can be argued that with large parts of the continent still having weak economies and infrastructure, Africa is a late comer into the economic development bandwagon. It has the potential of benefiting from new technologies that have been developed with sustainability in mind. Examples may include solar energy, more efficient irrigation systems, and better systems for storing and transporting food. Through these developments, WEF nexus securities can be attained sustainably, and their interdependencies better understood and harnessed.

Constraints: While rapid demographic growth offers labor for production and provides a market for goods and services, it also constitutes a significant challenge to achieving sustainable development outcomes (UNECA, 2013b). It ousts pressure on WEF nexus components by increasing demand and exploitation of both the resources, as well as the systems that support them. Africa's rapid urbanization also constitutes constraints to the achievement of security in WEF nexus components. Increased (especially unplanned or poorly planned) urbanization may bring about pressures on WEF nexus related products and services. About 30% of Africa's current population of about 1 billion people currently live in cities—a number that is forecast to grow within the next decades (UNECA, 2013b). Globalization may also be seen as a constraint for the attainment of securities in WEF nexus components. One of the most common characteristics of globalization in the African context has been the opening of local markets to goods and services from the developed world. On the other hand, different tools of market protectionism have been used by developed economies to limit access to African goods and services into their markets. Such protectionism has been seen to have an adverse effect on the economic development of poor countries. Low levels of economic development limit the ability of national governments to raise development capital, as well as contribute to limiting their capacity to manage key sectors of national social, economic and political life. Corruption feeds on these weaknesses and inefficiencies. Political instability has been an ongoing feature of the African landscape since the 1960s when most of the countries of the continent gained independence. Empirical studies have found that higher degrees of political instability are associated with lower levels of economic growth based on rates of GDP per capita (Aisen and Veiga, 2013). Political stability reduces people's potential to access WEF nexus products and services, as well as the ability for communities or countries to develop these resources. Land degradation continues to be a critical constraint for sustainable development in Africa. Its contribution to the decline of the continent's ability to sustain products and services to meet water, energy, and food securities are compounded by challenges posed by climate change.

The operating environment: Efforts that strive to achieve securities in WEF nexus components in Africa have to deal with a host of challenges at the local and national level. One of such challenges is the generally low levels of livelihood resources for many of the continent's populations (especially those in rural areas). In many cases, local participation in the management of local resources may not have been inclusive (especially when such resources are developed or exploited by non-local interests). UNECA (2013b) report that in Africa, culture and tradition dictate the choices made by communities regarding use and conservation of natural resources. Cultural norms and traditions may be associated with the perception, exploitation, and management of such resources, and need to be taken into consideration in related planning and policies. In some areas, local disease burdens may be affecting the health of WEF nexus components, as well as affecting the relationship of such components to local populations (such as the relationship between water, sanitation, and food).

Outcomes: Besides the numerous outcomes that can be derived from understanding and harnessing interdependencies in the WEF nexus, the socio-political, economic and environmental context in Africa points to some specific potential outcomes for the continent. The sustainable management of supplies in the face of increasing demand and exploitation pressures is required to ensure judicious management of WEF nexus components. Proper understanding and harnessing interdependencies in the WEF nexus will lay the groundwork for sustaining resilient, productive environments and ensuring intergenerational equity in access to these resources. This can also contribute to greater understanding in the management of conflict in transboundary resources and ensure human security.

Interaction of WEF Nexus Components in Africa

In sub-Saharan Africa, five main interactions contribute to cementing the interconnectedness in the WEF nexus: water for energy; water for food; energy for water; energy for food, and food for energy.

Water for energy

The use of water for the production of energy at different stages and in various forms (Gulati, 2014) constitutes the water for energy dimension of interactions within the WEF nexus. The utilization of water for hydroelectric power generation is an example in this dimension. Africa currently has a total of about 980 operational hydropower plants in about 34 countries, with an average capacity of 26 MW. About 392 of these plants are large dams with a capacity of 10 MW or more, 588 plants are small dams with a capacity of less than 10 MW (IEA, 2015; IRENA, 2012). These hydropower plants contribute to the social and economic well-being of many people. Water is also used to produce energy in indirect ways, such as for irrigating bioenergy feedstock. Since the World Food Price Crisis of 2008–09, the phenomenon of Large-Scale Land Acquisition (LSLA) for the cultivation of biofuel feedstock has expanded in SSA. The acquisition of land for this purpose has been closely linked to the acquisition and use of water resources to sustain the production of these biofuel crops. Some scholars have identified the "grab" of water as being at the heart of such land acquisitions (Mehta et al., 2012; Rulli et al., 2013). Negative outcomes of LSLA on the access of local peoples to land-based resources such as water, as well as on food security has been reported (Yengoh and Armah, 2014, 2015). The energy sector relies heavily on water for a variety of uses. Water is an effective medium for transporting quantities of waste heat generated by power plants. It is, therefore, an efficient resource for cooling in thermal power plants. It is estimated that about 580 billion cubic meters of freshwater are withdrawn for energy production annually. This accounts for 15% of the world's total water withdrawal. Freshwater use for power generation is expected to increase by 35% by 2035 (IEA, 2012). Given this scenario, competition for freshwater is likely to increase as food and energy production become ever thirstier for water. On the other hand, water availability is likely to decrease in many countries and regions of climate variability and change.

Water for food

In SSA, a majority of the food production is rain-fed. The FAO reports that less than 2% of the continent's agricultural area is equipped for irrigation (FAOSTAT, 2016). It has been observed that optimizing techniques of rainwater harvesting can increase yields up to three times. Besides providing more water for crops, rain water harvesting also has the potential of contributing to the recharge of groundwater, and to reduce soil erosion (FAO, 2008b). Just about 5% of cropland in the region is irrigated, even though parts of the region have largely unexploited surface water and groundwater resources that can be harnessed for the benefit of agricultural production (FAO, 2008b). The FAO (2008b) estimates that irrigated land in developing countries will increase by about 34% by 2030, while the amount of water used by agriculture will increase by about 14%. This will place additional pressure on, as well as competition for water resources between different sectors. There is a growing body of knowledge which asserts that hydro climatic variability is increasingly a problem for agriculturalists in SSA (Brown et al., 2011). Given the high reliance of societies and economies in the region on agriculture, this variability is contributing to challenges at a range of levels—from food production constraints and lack of employment opportunities at the local level to lower GDP, migration, and even potential for socio-political instability at the national level.

Energy for water

Energy is used in different forms and at different scales to harness water resources in SSA. For irrigation purposes, this energy may take the form of simple human powered pumps or small motorized pumps used by small and medium scale agriculturalists. These pumps are used to force water to the surface from relatively shallow underground reserves. Energy is also used for different processes associated with the treatment, distribution, and storage of water for domestic and industrial purposes. Large amounts of energy are also required to deal with the waste water resulting from municipal use, industrial processes, and other uses. This entails the use of water for the manufacture of chemicals utilized in the treatment of waste water, as well as associated treatment procedures. New forms of energy development such as hydraulic fracturing for natural gas, oil sands development, and thermal power generation are all important sources of energy that are increasingly taking prominent places in the world's energy landscape (Hardy et al., 2015). Large quantities of water are required in the production of these new sources of energy, and there are important quality issues associated with the discharges of used water from these processes into the natural environment.

Energy for food

Modern food production depends substantially on inputs of energy. Such energy inputs can be harnessed from a range of sources—from human manual labor and draught animal power to sophisticated high-tech machinery. Such labor inputs contribute to a wide variety of food production activities on farms (for example land preparation, fertilizer application, irrigation), in food production factories, in the transportation, and storage of food. In most of the developed world and parts of developing countries where agricultural intensification has replaced small-holder production systems, the practice of agriculture is very energy-intensive. In most of SSA, a lot of the farming practice is still subsistence (relying heavily on human manual labor), or semi-intensive at best (incorporating draught power, and small-scale machinery for some aspects of food production). Given the desire to obtain higher yields per unit area of ever decreasing land sizes, African farmers (irrespective of scale) will make use of opportunities of using inorganic fertilizers when and where possible. Inorganic fertilizers are energy-intensive inputs into crop cultivation. Farmers will also take advantage of possibilities of using small-scale machinery for a range of agricultural production activities where possible. Irrespective of the scale of production, there is a strong dependence of agriculture on energy. The food sector currently accounts for around 30% of the world's total power consumption (FAO, 2011). Given the close relationship between food production and energy use, it is important always to factor the implications of energy demand, and supply when planning food production and security scenarios. Global scenarios of increased food production required to meet future demand, therefore, have an energy component that is often not explicitly reported. It is estimated that an increase in food production of 60% will lead to an increase in energy consumption in agriculture of 84% (Pimentel and Pimentel, 2007).

Food for energy

The food for energy interaction in the WEF nexus is not immediately evident. In SSA, the phenomenon of LSLA offers strong insights into this relationship. When large quantities of small-holder farmlands and lands on which local peoples depend on for non-farm food resources is converted to biofuel production, priorities for the provision of food are replaced with those of bioenergy. Besides farmed food crops, a range of land resources are usually lost as a result of these large-scale investments in bioenergy feedstock. These include non-farmed land on which local populations depend for edible wild products—including both land-based and aquatic resources (Yengoh and Armah, 2014).

Another food for energy scenario can be understood by examining food wastes. This is a more global interaction, argued to be more common in the developed world than in developing countries. UNEP and the World Resources Institute (WRI) estimate that one-third of all food produced worldwide gets lost or wasted in the production and consumption systems. This includes losses during production and harvesting, processing and packaging, handling and storage, distribution and marketing, as well as consumption (Lipinski et al., 2013). Consumers in the developed world waste about 22 million tons of food, close to the net food production of SSA, 230 million tons (Lipinski et al., 2013). In 2009, SSA contributed to 9% of global food losses, compared to 6% in Latin America, 7% in North Africa, 23–28% in Asia and 14% in North America as well as in Europe. A majority of the food wastes in sub-Saharan Africa and most other developing countries are a result of poor processes of harvesting, transportation, and storage. In the developed world, most of the food wastes generated arise from patterns of shopping and consumption. The energy implications of these food losses are two-fold—the loss of energy in producing these foods which may have been put into alternative uses, or not produced and used at all; and the problem of dealing with the energy embodied in these foods which will contribute to GHG emissions into the atmosphere.

Challenges of Implementing the WEF Nexus Approach

The challenge of a changing climate

Sub-Saharan Africa has been described as one of the regions that is going to experience severe impacts of climate change (IPCC, 2014; UNFCCC, 2007). For some reasons, the region's WEF nexus components stand to be negatively affected substantially by climate variability and change. The agricultural sector is an important economic activity for almost all countries of the region, in some cases constituting a large share of national GDP. This sector is however closely tied to the patterns of rainfall as most farming is rain-fed. The agricultural sector, therefore, stands to suffer from changes in the patterns, frequency, and quantity of precipitation forecasted for some countries in the region (de Fraiture et al., 2007; Müller et al., 2011; UNFCCC, 2007). In the same vein, many SSA countries largely rely heavily on hydropower for energy production. According to the IEA (2015) many of the over 30 countries in the region have very limited alternatives to hydroelectric power sources. This means that national energy security is highly dependent on river flows. High variability in river flows has been identified as one of the potential impacts of climate change in the region, and this stands to affect energy production as well as its associated supporting services. The adaptive capacity of many SSA countries to these climate change challenges is low (de Fraiture et al., 2007; UNFCCC, 2007). Climate variability and change, therefore, stands to pose serious challenges to water, food and energy security in future.

Demographic growth

Africa's fast-growing population is expected to double by 2050, contributing to significant stress on many of its natural resources (UNECA, 2013b). Global food demand, for example, is projected to increase by 60% in 2050 over the 2005–2007 base (Alexandratos and Bruinsma, 2012). Population growth is forecast to contribute 70% to this increase while changes in per capita calorie intake and change in diets will contribute 30%. The bulk of agricultural production increases are expected to happen in developing countries where most of the population growth and food demand increases are projected to occur. Alexandratos and Bruinsma (2012) estimates that 77% of the global annual agricultural production increase needs to occur in developing countries. A growing population may mean an increase in the market for goods and services, a potential for stimulating economic growth. However, to meaningfully contribute to the achievement of sustainable development goals Africa needs a population base that is well educated, and with the right skills mix (UNECA, 2013b).

Rapid urbanization

Rapid population growth and high rates of urbanization go hand-in-hand and are features of the current African demographic landscape. Most studies suggest that urbanization tends to have a positive impact on economic growth. Urban environments offer opportunities for education, employment, and health services which determine the development of new technologies and adoption of technologies, as well as the consumption of goods and services that can positively influence economic activity directly (Arouri et al., 2014). Africa's rate of urbanization has remained high over the last two decades. The growth of urban areas has however not been matched with an increase in the capacity of such centers to provide the necessary services required to harness the demographic potential resulting from rural-urban migration and high rates of fertility.

Urban centers constitute concentrated areas of strong demand for products and services such as water, energy, and food. To attain securities in WEF nexus components in such areas of high concentration of demand, efficient systems of production, transportation, and delivery of such services need to be available. This is lacking in many urban areas of the African continent. The result is an inequality in the availability and access to WEF nexus products and services, leading to deprivation across socioeconomic (rich versus poor) status and geographical (slum versus upscale dwellers) divides. Arouri et al. (2014) report that while African cities generate about 55–60% of the continent's GDP, about 43% of urban populations live below the poverty line.

Large-scale land and acquisitions for biofuel feedstock

In recent years, global demand for biofuels (biodiesel and ethanol) has been one of the main challenges of food production in sub-Saharan Africa. This challenge has taken the form of competition for land between large-scale biofuel investors and smallholder food producers in rural African communities. Global demand for biofuels is expected to increase threefold by 2050 (IEA, 2013). If current trends of biofuel land conversions continue, the amount of land that would be used for biofuel production (currently about 1% of the global arable land) could increase up to 3% by 2030 (FAO, 2008a). While this may also depend on sources of biofuel and the development of 2nd generation biofuels, such increases and pressure on arable lands are bound to have implications for food production, especially for rural small-holder agriculturalists.

Environmental degradation

Recent decades have seen substantial changes to the natural environment of the African continent. According to the United Nations Environmental Programme (UNEP), water resources are continuously affected by persistent droughts and land-use changes. Physical and chemical degradation is affecting up to 65% of farmland; habitat destruction, poaching, and rapid demographic growth is leading to the destruction of the continent's biodiversity as well as its loss (UNEP, 2008). Climate change stands to compound these challenges and to accelerate some of the negative outcomes that they impose to sustainable development in the region. These changes have the potential of undermining the ecological foundations that underpin the WEF nexus in Africa.

The nature of agricultural production

One of the main contributors to these changes is the nature of practice of Africa's main socio-economic activity—agriculture. Intensification, mechanization and labor productivity in SSA remains very low (with few exceptions), and yield gaps remain very significant for most major crops. Increase in agricultural production by increasing land area has gone hand in hand with deforestation, biodiversity loss and environmental degradation (UNEP, 2008). Besides the negative environmental outcomes associated with opening up new lands (especially prime forests) for agriculture, there are intergenerational concerns about the availability of land for future generations to meet their own land-based needs. Africa's land resource base has shrunk steadily over the last half-century. According to UNEP (2008) the per capita land base shrunk from 8.3 ha/person in 1970 to 3.2 ha/person in 2005, and is forecast to decline to 1.5 ha/person in 2050.

African Development and the WEF Nexus

Design projects to harness multiple benefits

Development projects that have the vision of harnessing natural resources for the greater local, national, or regional good have the potential of disrupting the interrelations between components of the WEF nexus. In some cases, such disruptions may be negative, and with far reaching outcomes for both current and future generations and across geopolitical regions. The development of an energy project such as the building of a large-scale hydroelectric dam upstream could reduce access to water for wetland dependent small-holder farming communities downstream—robbing them of food resources such as access to domestic water, navigation, rice production potential, and fisheries. In the same light, large-scale investments in bioenergy feedstock that takes away agricultural land from local communities could lead to food insecurity and limited access to water and associated water resources for communities that host such investments. Outcomes of such development projects do not necessarily need to be negative. Given appropriate consideration of the interaction of nexus elements, vital safeguards can be put in place to ensure that the outcomes of development projects safeguard community benefits related to all three nexus security dimensions. Hydroelectric dams, for example, could be multi purpose, so that besides generating renewable energy resources, they can produce associated benefits such as flood control, irrigation, and water supply. The three dams planned for the Niger Basin is an example of such a multi purpose planning that addresses broad demands of societal benefits from the WEF nexus (Bach et al., 2012). The project will lead to increased water flow, stabilized to be available in the dry season, and enabling two farming seasons as well as improved navigation (Bach et al., 2012). The dam project will lead to a five-fold increase in irrigated land area, create jobs and substantially improve self-sufficiency in rice, one of the key food items in this region (Bach et al., 2012).

Plan for inter-sectoral competition for key resources

The future of Africa's development is likely to be marked by substantial competition for resources between different sectors of socioeconomic and environmental growth. Being the most important socio-economic activity in the continent, agriculture's future water demand offers insights into the future of this competition. The rate of increase of the global area equipped for irrigation will expand by 6.6% by 2050 over the base 2005/2007. Most of this increase is expected to occur in the developing world—Africa, and East and South Asia. It is forecast that by 2050, irrigated agriculture will be practised on 16% of the total global cultivated area, and will be contributing to about 44% of total crop production (Alexandratos and Bruinsma, 2012). Increase in water demand for irrigation will compound competition over water resources for other key sectors which are expected to also witness increases over this period. Steep increases in domestic and industrial water use are expected. Such competition over water resources may lead to as much as an 18% decrease in water availability for agriculture by 2050 globally (Strzepek and Boehlert, 2010).

Support transboundary cooperation in the management of resources

There are many cases when the management of WEF nexus interactions and outcomes goes beyond local, national or even regional administrative regions. This can be exemplified by the administration of large river bodies which flow across several countries and support a range of services beyond the provision of water, energy, and food resources. In such cases, transboundary resource management procedures become essential to ensure equitable access and the sustainability of such resources. A number of management commissions are tasked with meeting these goals such as the Lake Chad Basin Commission (LCBC), the Permanent Okavango River Basin Water Commission (OKACOM), the Niger Basin Authority, the Zambezi Watercourse Commission, and the Nile River Basin Initiative. At the heart of these initiatives and communions is the need to strengthen cooperative management among member states to sustain systems that support the production of water, energy, and food, with the ultimate goal of facilitating sustainable and inclusive growth, climate resilience, and poverty reduction.

Support policies that contribute to building greener economies

Given the importance of WEF nexus components on the socio-economic development of the continent, it may be useful for African leaders and governments to include water energy and food governance as priority sectors for integrated development. This may call for the establishment of appropriate policy and legal instruments that will ensure long-term engagement in understanding, developing and sustaining the basis for resilience in WEF nexus sectors. The role of ecosystem services is therefore essential. Ecosystem services support the provision of water, food, and energy and are vital to the attainment of securities in these sectors. Such services, therefore, form an essential part of the foundation of a green economy. Examples of policies that strengthen the basis for resilience of WEF nexus components and enhances benefits from WEF nexus interdependencies may include:

- Developing and allocating resources (physical, technical, and human) to explore and strengthen the linkages between WEF nexus sectors and support initiatives that improve cross-sectoral collaborative mechanisms and institutions.
- Strengthening the development and use of data and decision support systems for dialogue across all levels of decision-making.
- Providing support for institutional reform and capacity building to relevant institutions and organizations (national and regional), as well as processes of knowledge transfer to ensure a lasting culture of integrating WEF nexus into the national and regional development agenda.
- Developing frameworks which ensure that the development programs in each of the nexus sectors (water, energy, and food) consider outcomes and stakeholders of other sectors in an integrated manner.

Conclusion

Adequate access to water, energy, and food of good quality and sufficient quantity in Africa remains low despite the continent's rich natural resource base. The problems that contribute to this inadequacy of access are social, economic, political, and even cultural. Other challenges such as climate change and land degradation do also contribute to this limited access and are forecast to become increasingly influential in the future. There is, therefore, a crucial need to properly understand the functioning of the water, energy and food systems on the continent. The WEF nexus provides a basis for understanding the interconnectedness of these three resource sectors and a framework for the development of policies and strategies for their sustainable management. Africa's rich natural resource base can form the foundation for development and management of WEF nexus components and for achieving sustainable green growth. Other factors that may contribute to this end are the region's growing consumer base and its abundant labor. A number of hurdles will, however, need to be addressed. These include developing adaptive governance that promotes flexibility, cooperation and collaboration to cope with the challenges of integrated development across WEF nexus sectors.

References

AICD. 2017. Africa - Electricity Transmission and Distribution Grid Map. Africa Infrastructure Country Diagnostic (AICD), Africa Electricity Transmission Network. The World Bank Group. Washington DC.

Adams, S. and Opoku, E.E.O. 2016. Population growth and urbanization in Africa: implications for the environment. *Population Growth and Rapid Urbanization in the Developing World* : IGI Global pp. 282–297.

Aisen, A. and Veiga, F.J. 2013. How does political instability affect economic growth? *European Journal of Political Economy* 29: 151–167.

Alexandratos, N. and Bruinsma, J. 2012. World agriculture towards 2030/2050: the 2012 revision: ESA Working paper Rome, FAO.

Arouri, M., Youssef, A.B., Nguyen-Viet, C. and Soucat, A. 2014. Effects of urbanization on economic growth and human capital formation in Africa (Vol. PGDA Working Paper No. 119, pp. 22). Massachusetts, USA: Program on the Global Demography of Aging, Havard University. http://www.hsph.harvard.edu/pgda/working.htm.

Bach, H., Bird, J., Clausen, T.J., Jensen, K.M., Lange, R.B., Taylor, R., . . . Wolf, A. 2012. Transboundary River Basin Management: Addressing Water, Energy and Food Securiy: Mekong River Commission, Lao PDR.

Bazilian, M., Rogner, H., Howells, M., Hermann, S., Arent, D., Gielen, D., . . . Tol, R.S. 2011. Considering the energy, water and food nexus: Towards an integrated modelling approach. *Energy Policy* 39(12): 7896–7906.

Bizikova, L., Roy, D., Swanson, D., Venema, H.D. and McCandless, M. 2013. The water–energy–food security nexus: Towards a practical planning and decision-support framework for landscape investment and risk management. *IISD Report* (pp. 28). Manitoba, Canada: International Development Research Centre (IDRC).

Brauman, K.A., Richter, B.D., Postel, S., Malsy, M. and Flörke, M. 2016. Water depletion: An improved metric for incorporating seasonal and dry-year water scarcity into water risk assessments. *Elementa, 4.*

Brown, C., Meeks, R., Hunu, K. and Yu, W. 2011. Hydroclimate risk to economic growth in sub-Saharan Africa. *Climatic Change* 106(4): 621–647.

Colding, J. and Folke, C. 2001. Social taboos: "invisible" systems of local resource management and biological conservation. *Ecological Applications* 11(2): 584–600.

David, L. Tschirley, Jason Snyder, Michael Dolislager, Thomas Reardon, Steven Haggblade, Joseph Goeb, Lulama Traub, Francis Ejobi, Ferdi Meyer. 2015. Africa's unfolding diet transformation: implications for agrifood system employment. Journal of Agribusiness in Developing and Emerging Economies, 5(2): 102–136, https://doi.org/10.1108/JADEE-01-2015-0003.

de Fraiture, C., Smakhtin, V., Bossio, D., McCornick, P., Hoanh, C., Noble, A., . . . Finlayson, M. 2007. Facing climate change by securing water for food, livelihoods and ecosystems. *Journal of Semi-arid Tropical Agricultural Research,* 4(1).

ESCWA. 2015. Conceptual frameworks for understanding the water, energy and food security nexus. *Working Paper* (Vol. E/ESCWA/SDPD/2015/WP, pp. 27). Beirut, Lebanon: United Nations Economic and Social Commission for Western Asia (ESCWA).

FAO. 2008a. Climate change, biofuels and land Info Sheet for High Level Conference. Rome, Italy: Food and Agriculture Organization of the United Nations Organization (FAO) ftp://ftp.fao.org/nr/HLCinfo/Land-Infosheet-En.pdf.

FAO. 2008b. Water at a Glance. The relationship between water, agriculture, food security and poverty (pp. 15). Rome Italy: Water Development and Management Unit, Food and Agriculture Organization of the United Nations. http://www.fao.org/nr/water/docs/waterataglance.pdf.

FAO. 2011. Energy-Smart Food for People and Climate *Issue Paper* (pp. 78). Rome, Italy: Food and Agriculture Organization of the United Nations (FAO).http://www.fao.org/docrep/014/i2454e/i2454e00.pdf.

FAO. 2014. The Water-Energy-Food Nexus: A new approach in support of food security and sustainable agriculture (pp. 28). Rome Italy: Food and Agriculture Organization of the United Nations (FAO).

FAO. 2015. Regional Overview of Food Insecurity: African food insecurity prospects brighter than ever (pp. 35). Accra, Ghana: Food and Agriculture Organization of the United Nations (FAO).

FAO, IFAD and WFP. 2015. The State of Food Insecurity in the World 2015: Meeting the 2015 international hunger targets: taking stock of uneven progress (pp. 62). Rome, Italy: Food and Agriculture Organization of the United Nations (FAO), the International Fund for Agricultural Development (IFAD) or of the World Food Programme (WFP).

FAOSTAT. 2016. FAO Statistics Database: Statistics Division, United Nations Food and Agriculture Organization (FAO). Consulted on the 27/06/2016: http://faostat3.fao.org/home/E.

Faurès, J.-M. and Santini, G. 2008. Water and the rural poor: interventions for improving livelihoods in sub-Saharan Africa. *Enabling Poor Rural People to Overcome Poverty* (pp. 109). Rome, Italy: United Nations Food and Agriculture Organization (FAO).

Foley, J.A., Ramankutty, N., Brauman, K.A., Cassidy, E.S., Gerber, J.S., Johnston, M., . . . West, P.C. 2011. Solutions for a cultivated planet. *Nature* 478(7369): 337–342.

Fox, S. 2012. Urbanization as a global historical process: Theory and evidence from sub-Saharan Africa. *Population and Development Review* 38(2): 285–310.

Gulati, M. 2014. Through the energy and water lens *Understanding the Food Energy Water Nexus* (pp. 28). Gland, Switzerland: World Wildlife Fund (WWF).

Handley, A. 2015. Varieties of capitalists? The middle–class, private sector and economic outcomes in Africa. *Journal of International Development* 27(5): 609–627.

Hardy, D., Cubillo, F., Han, M. and Li, H. 2015. Alternative water resources: A review of concepts, solutions and experiences (pp. 71). The Hague, The Netherlands: Alternative Water Resources Cluster, International Water Association (IWA).

Hoff, H. 2011. Understanding the Nexus. Background Paper for the Bonn. 2011 Conference: The Water, Energy and Food Security Nexus (pp. 51). Stockholm, Sweden: Stockholm Environment Institute, Stockholm (SEI).

Howarth, C. and Monasterolo, I. 2016. Understanding barriers to decision making in the UK energy-food-water nexus: The added value of interdisciplinary approaches. *Environmental Science & Policy* 61: 53–60.

IEA. 2012. Water for energy: Is energy becoming a thirstier resource? *World Energy Outlook 2012* (pp. 12). Paris, France: International Energy Agency (IEA).

IEA. 2013. World Energy Outlook 2013 (pp. 708). Paris, France: International Energy Agency.

IEA. 2015. World Energy Outlook 2015—Excecutive Summary (pp. 12). Paris, France: International Energy Agency (IEA).

IPCC. 2014. Climate Change 2014—Impacts, Adaptation and Vulnerability: Regional Aspects. Cambridge University Press. Cambridge, UK: Intergovernmental Panel on Climate Change.

IRENA. 2012. Prospects for the African Power Sector: Scenarios and strategies for Africa Project (pp. 60). Abu Dhabi, United Arab Emirates: International Renewable Energy Agency.

Karabulut, A., Egoh, B.N., Lanzanova, D., Grizzetti, B., Bidoglio, G., Pagliero, L., . . . Maes, J. 2015. Mapping water provisioning services to support the ecosystem–water–food–energy nexus in the Danube river basin. *Ecosystem Services.*

Keairns, D., Darton, R. and Irabien, A. 2016. Water-energy-food nexus. *Annual Review of Chemical and Biomolecular Engineering, 7*(1).

Lipinski, B., Hanson, C., Lomax, J., Kitinoja, L., Waite, R. and Searchinger, T. 2013. Reducing food loss and waste. *World Resources Institute Working Paper, June.*

Mehta, L., Veldwisch, G.J. and Franco, J. 2012. Introduction to the Special Issue: Water grabbing? Focus on the (re)appropriation of finite water resources. *Water Alternatives* 5(2): 193.

Mueller, N.D., Gerber, J.S., Johnston, M., Ray, D.K., Ramankutty, N. and Foley, J.A. 2012. Closing yield gaps through nutrient and water management. *Nature* 490(7419): 254–257.

Müller, C., Cramer, W., Hare, W.L. and Lotze-Campen, H. 2011. Climate change risks for African agriculture. *Proceedings of the National Academy of Sciences* 108(11): 4313–4315.

Pimentel, D. and Pimentel, M.H. 2007. Food, Energy, and Society: CRC press.

Rulli, M.C., Saviori, A. and D'Odorico, P. 2013. Global land and water grabbing. *Proceedings of the National Academy of Sciences* 110(3): 892–897.

Smajgl, A., Ward, J. and Pluschke, L. 2016. The water–food–energy Nexus–Realising a new paradigm. *Journal of Hydrology* 533: 533–540.

Strzepek, K. and Boehlert, B. 2010. Competition for water for the food system. *Philosophical Transactions of the Royal Society B: Biological Sciences* 365(1554): 2927–2940.

UN-DESA. 2015. World Population Prospects: The 2015 Revision (pp. 60). New York, USA: United Nations, Department of Economic and Social Affairs, Population Division. https://esa.un.org/unpd/wpp/publications/files/key_findings_wpp_2015.pdf.

UNDP. Human Development Report 2015: Work for Human Development. pp. 288. *In*: Jahan, S. (ed.). New York, USA: United Nations Development Programme (UNDP). http://hdr.undp.org/sites/default/files/2015_human_development_report.pdf.

UNECA. 2013a. Making the Most of Africa's Commodities: Industrializing for Growth, Jobs and Economic Transformation (pp. 260). Addis Ababa, Ethiopia: United Nations Economic Commission for Africa (UNECA).

UNECA. 2013b. Managing Africa's Natural Resource Base for Sustainable Growth and Development (pp. 216). Addis Ababa, Ethiopia: United Nations Economic Commission for Africa (UNECA).

UNEP. 2008. Atlas of Our Changing Environment (pp. 393). Nairobi, Kenya: Division of Early Warning and Assessment (DEWA), United Nations Environmental Programme (UNEP).

UNFCCC. 2007. Climate change: impacts, vulnerabilities and adaptation in developing countries (pp. 68). Bonn, Germany: United Nations Framework Convention on Climate Change (UNFCCC).

UNICEF and WHO. 2015. 25 Years Progress on Sanitation and Drinking Water: 2015 Update and MDG Assessment (pp. 90). Geneva, Switzerland: United Nations Children's Emergency Fund, and World Health Organization. http://www.publicfinanceforwash.com/resources/25-years-progress-sanitation-and-drinking-water-2015-update-and-mdg-assessment.

United Nations Organisation. 2010. Summary Report and Recommendations, UN Secretary General's Advisory Group on Energy and Climate Change (AGECC), 28 April 2010, p. 13. http://www.un.org/chinese/millenniumgoals/pdf/AGECCsummaryreport%5B1%5D.pdf.

UNO. 2015. World Population Prospects. Key Findings and Advanced Tables. The 2015 Revision (Vol. ESA/P/WP. 241, pp. 66). New York, USA: Department of Economic and Social Affairs Population Division, United Nations Organization.

World Bank. 2015. World Bank Data. Washington DC, USA: The World Bank. Accessed on 07/06/2016: http://www.worldbank.org/en/topic/energy/overview#1.

World Bank. 2016. Fact sheet: The World Bank and energy in Africa. Washington DC, USA: The World Bank.

Yengoh, G.T. and Ardö, J. 2014. Crop yield gaps in Cameroon. *Ambio* 43(2): 175–190.

Yengoh, G.T. and Armah, F.A. 2014. Land access constraints for communities affected by large-scale land acquisition in Southern Sierra Leone. *GeoJournal*, 1–20.

Yengoh, G.T. and Armah, F.A. 2015. Effects of large-scale acquisition on food insecurity in Sierra Leone. *Sustainability* 7(7): 9505–9539.

Chapter-5

The Water-Energy-Food Nexus in Europe

Fabiani S.,[1] *Dalla Marta A.,*[2] *Orlandini S.,*[2] *Cimino O.,*[1] *Bonati G.*[1] and *Altobelli F.*[1,*]

INTRODUCTION

This chapter refers to the definition of a new concept of sustainability assessment, the Water-Energy-Food nexus approach. It deals with the linkages and interactions between water and energy management policies, related to food production and allow to gain a global and integrated view to improve agriculture sustainability. Within the chapter we'll focus on water and energy regulatory framework mainly regarding agricultural sector, to highlight synergies and opportunities of the nexus perspective.

Understanding the Nexus at EU Level (Main Constraints for Sustainable Agriculture)

The sustainability of natural resources management, under economic, environmental and social perspective, needs to be assessed based on a fair balance between the use and the availability of resources.

In agriculture, such an assessment must consider the main constraints of global trends: expected growth of population (impact of food production), decreasing availability of natural resources such as soil and water (productive input), and increasing GHGs emissions (related to climate change). This implies a deep transformation of the current model of development, that should be able to describe and address the complex and interrelated nature of global resources looking at "Water-Energy-Food *nexus*" (WEF) as a conceptual framework.

[1] CREA PB - Research Centre for Policy and Bioeconomy; Council for Agricultural Research and Economics, Via Po 14 00198 Roma (Italia).
[2] DiSPAA – Department of Agrifood Production and Environmental Sciences, University of Florence, Piazzale delle Cascine 18 - 50144, Firenze (Italia).
* Corresponding author: filiberto.altobelli@crea.gov.it

This approach suggests that sustainability driven strategies at all levels (global to regional to local), must be planned and implemented considering and assessing the linkages between water uses and availability, energy consumption and sources, land use and food production. Stakeholders play an important role in assessing the WEF *nexus*, as many policy choices affect their possibility to make use of environmental goods and ecosystem services (FAO, 2014).

For assessing the WEF *nexus*, we need to consider different aspects. On one hand, that **food production** needs different kind of inputs: human (labor, time, management/administration), socio-economic (money, governance institutions, participation), natural (sunlight, soil, water, wind, petroleum, gas), and human-made (electricity, alternative water, machinery, agrochemicals) resources.

On the other, that agriculture uses **water** from different sources: "green water", the rain infiltrated into the soil and available to crops (Falkenmark, 2003), and "blue water" including surface water from rivers and lakes and groundwater from aquifers. To make blue water and water generated from alternative sources (desalination, re-use) available to crops, it is necessary to spend energy. Both direct (fuel consumption, electricity for pumping, etc.) and indirect (fertilizers, pesticides, seeds, etc.) energy used in agriculture requires again water.

In particular, irrigated agriculture is the most resource-intensive form of food production causing also significant diffuse pollution of soil and groundwater. The sustainable agriculture goal is thus to implement innovative tools and service capacities able to optimize input management (energy, nutrients and water) and productivity of intensive systems, with the vision of bridging sustainable crop production with fair economic competitiveness.

At EU level in a business as usual scenario by 2030 water, energy and food demand is expected to increase by 30–50% (US National Intelligence Council, 2012). Based on this projection, achieving universal access to food, drinking water and modern energy within the "planetary boundaries" can be considered the main challenge of our society. This requires considering the interconnections, the vulnerabilities and the finiteness of natural resources within the policy and regulatory frameworks, the governmental planning mechanisms, the production processes and the consumption patterns.

Water-Energy-Food Nexus as a Concept to Describe and Address the Complex and Interrelated Nature of Global Resources

WEF *nexus* is a conceptual approach developed to better understand and systematically analyze the interactions between the natural environment and human activities, and to work towards a more coordinated management and use of natural resources across sectors and scales. This can contribute to identify and manage trade-offs and to build synergies through our responses, allowing for more integrated and cost-effective planning, decision-making, implementation, monitoring and evaluation (FAO, 2014).

The Bonn Conference, organized in preparation for the Rio+20 Summit, provided evidence that improved water, energy and food security can be achieved through a nexus approach. The nexus approach can also support the transition towards a Green

Economy, which aims, among other things, at resource use efficiency and greater policy coherence (Hoff, 2011).

In particular, the nexus approach is becoming a key tool to address and manage some of the major challenges that our society and its rapid transformations are posing. Among them: (I) Scarcity of water and other natural resources, (II) Climate change, (III) Degradation of the resource base, and (IV) Water, energy and food security.

Scarcity of water, land and other resources. Scarcity of resources is rapidly escalating due to increasing demand, resource degradation and pollution. By 2050 agricultural production would have to grow by another 70%, and agricultural land would have to expand by about 10% globally, by 20% in developing countries and by 30% in Latin America (Bruinsma, 2009). Even the most optimistic scenarios of improvements in productivity through technological development, still project an increase in agricultural water demand of at least 20% by 2050 (De Fraiture et al., 2007).

Climate change. Climate change and variability add further pressures, by accelerating degradation of drylands, reducing glacier water storage, increasing frequency and intensity of extreme events (such as droughts or floods), and decreasing reliable water supplies, as well as reliable and stable agricultural productivity. At the same time climate adaptation measures, such as intensified irrigation or water desalination, are often energy intensive.

Degradation of the resource base. Growing demand of natural resources and their unsustainable management have increased human ecological footprint and caused degradation of the natural resource base in many regions, including severe modification of ecosystems.

Water, energy and food security. Resource limitations in all sectors require moving towards increased resource use efficiency, demand management and more sustainable consumption patterns (http://www.wrc.org.za/News/Pages/Internationalandlocalwaterspecialistsconvenetotacklefreshwatergovernanceissues. aspx).

Synergies Between Water, Energy and Agriculture in the European Policy Framework

Energy policies

Over the past two decades, reducing energy consumption, improving energy efficiency and promoting production from renewable sources have become more and more important. Energy efficiency is at the heart of the European strategies to boost the transition to a resource efficient economy, cardinal principle of Europe 2020 strategy. The reason is that it is considered one of the most cost effective ways to enhance security of energy supply, and to reduce emissions of greenhouse gases and other pollutants, thus contributing to environmental objectives of mitigating climate change. This strategy can also help improving productivity and competitiveness for EU, especially in the face of increasing competition from China and North America. Europe is definitely boosting energy policies, and it is shown by the fact that in the

"Climate Energy pack" (known as EU 20-20-20 goal) the first target set for 2020 of primary energy saving of 20% through energy efficiency—a key step towards achieving our long-term energy and climate goals—has been empowered in Europe's 2030 strategy, setting up an indicative target at the EU level of at least 32%.

If energy targets will be achieved, economic analysts talk about 60 billion Euro on imports of oil and gas that could be saved by 2020, fundamental goal both for energy and economic security. Also, it is expected that achieving the target for renewable energy production will create over 600,000 jobs in the EU, by boosting innovation and technology sectors.

Renewable energies sector is a typical example of this process, driving the green growth of EU economies, reducing environmental impacts (lowering CO_2 emission) and generating positive socio-economic effects, mainly on occupation. Such sectors particularly affect agriculture in a positive way, allowing farmers to re-use waste and optimize sub-products for bioenergy production. In this framework, agricultural policies played a key role in supporting farmers.

Agricultural policies

The second pillar of the Common Agriculture Policy (CAP) provides specific measures, included in national Rural Development Plans (RDP), supporting the production and use of renewable energy in agriculture as well as energy efficiency (for instance investments in renewable energy production systems, advisory expenses for optimizing energy consumption, etc.).

Although CAP does not provide direct support for the production of biomass for bioenergy, EU makes available a set of instruments to boost bioenergy production, ranging from investments in physical assets, to the support of basic services and village renewal in rural areas, to measures aimed at helping actors in the agriculture and forestry sectors to work together, such as farmers, forest owners and business organizations.

As a result, in 2010 bioenergy and biofuels from agriculture and forestry contributed to around 63% of renewable energy generation in the EU-27 for a total amount of around 2600 PJ (petajoules),[1] confirming their crucial role to reach the renewable energy targets, which is why the EU member states incorporated the bioenergy option in their National Renewable Energy Action Plans (NREAPs).

Nonetheless, accurate assessment of the environmental sustainability, evaluation of the technological developments, calculations of direct and indirect emissions from biofuels, land use change and different bioenergy pathways are the crucial challenges of the "agriculture for energy" framework. All costs and benefits of biofuels production and use have to be taken into account to ensure a well-balanced policy. Resource assessment is therefore of particular importance. Knowing how much resource is available, at what cost and with which environmental impact, gives indeed a better insight into some environmental parameters such as water needs, soil degradation and the impact on biodiversity.

---

[1] As set in the Climate Energy pack, the EU 20-20-20 target foresee also that 10% of the transport sector final energy consumption should come from renewable energy sources, which is mostly provided in the form of liquid biofuels.

In Indirect Land Use Change (ILUC) framework, where sustainability plays a key role, the analysis of the water footprint for bioenergy production is considered a crucial issue, adding to the "agriculture for energy" framework the strategic role of water.

More generally, at global scale, water is strictly linked with energy production also because "traditional" thermo-electric plants require massive amounts of water for cooling, that's why reduction of energy consumption as well as use of alternative energy sources must be well connected with water policies, mainly the Water Framework Directive (WFD) 2000/60/UE.

Water policies

The European water legislation has been deeply transformed with the WFD that enacted in 2000 and went into full operation by 2012 (CEC, 2000), becoming one of the main environmental EU legislations. It responds to the environmental goals of the EU SDS (Sustainable Development Strategy) in relation to water use and water ecosystems protection. Its objective is to achieve the good ecological status of all water bodies in the EU for 2027, and maintain and promote sustainable water use in a long-term perspective. It requires securing environmental flows in rivers and recharge levels in aquifers.

On the other hand, the WFD socio-economic component that seeks to enforce cost-recovery for all water services can strongly affect the agricultural sector. In fact, if the WFD is fully implemented to recover all water-related costs, water tariffs will raise considerably causing water use reductions and inflicting substantial losses to farm income. Thus, the application of the WFD might question the viability of certain irrigated areas in many European regions.

In addition to the WFD, the EU Communication on Water Scarcity and Drought (CEC, 2007) was launched to stress the need for a full implementation of the WFD, and especially the application of effective water pricing and water efficient technologies as a way to improve water management. It highlights the need for improving drought risk management, fostering water-saving technologies and practices in all sectors. It especially underlines the integration of water-related issues in all sectorial policies such as the CAP, and reinforces the water saving potential of the irrigation sector in the EU that can mount up to 43% of the total water volume abstracted (Dworak et al., 2007).

Conclusion

The Water, Energy and Food (WEF) nexus means that the water security, energy security and food security are linked and interact with each other and with the environment. Under this approach, the sustainability of natural resources management must be assessed by taking into account a fair balance between uses and availability of those resources.

In agriculture, such an assessment must consider the main constraints of global trends, as with the growth of global population, consumption of water, energy, and food will also increase, placing stresses on these three sectors, and raising the importance of the WEF.

Many sustainable management assessments have been carried out at all levels. Naturally, their complexity increases with geographical scale, timescale and number of systems considered. Assessments can help identify trade-offs between the WEF systems and to design strategies to mitigate possible drawbacks. In particular, policies and planning related to water, energy and food resources should prioritize the long-term security of systems, because global projections of increasing population, urbanization and changes in consumption patterns suggest increasing future demands on WEF systems. In fact, only for the European Union in a business as usual scenario by 2030 water, energy and food demand is expected to increase by 30–50%.

The Food and Agriculture Organization (FAO) defined food security as: "…when all people, at all times, have physical, social and economic access to sufficient, safe and nutritious food which meets their dietary needs and food preferences for an active and healthy life" (World Food Summit, 1996). The Global Water Partnership (GWP) defined water security as: "Ensuring the availability of adequate and reliable water resources of acceptable quality, to underpin water service provision for all social and economic activities in a manner that is environmentally sustainable; mitigating water-related risks, such as floods, droughts and pollution…" (GWP, 2012). Finally, energy security has been defined as the uninterrupted availability of energy sources at affordable prices. There is an evolving agenda around the promotion of diversity, efficiency and flexibility within the energy sectors so that they are able to respond to energy emergencies. However, 87% of the energy supply comes from oil, coal and gas. Agricultural energy use increases in the course of agricultural intensification, both directly (i.e., through the use of energy intensive fertilizers, tractors and the pumping of water), and indirectly (i.e., transportation, processing, packing, refrigeration) (Lele et al., 2013).

Although in such complex and unpredictable systems water, food and energy security can never be fully achieved, understanding the interactions in WEF systems can help decision makers balancing trade-offs, in response to an increasing demand. The relationship between these systems, in fact, goes beyond simple interdependence of resources (i.e., how demand for one resource can drive demand for another), and a deeper analysis is required for a more effective policy planning in the three sectors.

In terms of energy policy, the cardinal principle of Europe 2020 strategy is reducing energy consumption, improving energy efficiency and promoting production from renewable sources.

In this context, a key role is played by the development and spread of innovative technological solutions to achieve emission reduction commitments set in the sectorial EU policies. In particular, in international negotiations Europe has proposed a GHG reduction of 20% in the developed countries by 2020. One of the principal tools to achieve this target is to apply the Energy Efficiency Action Plan. In this sense, the amount of energy produced from renewable sources plays an important role.

Renewable energy is, in fact a sector driving the green growth of EU economies, reducing CO_2 emissions and generating positive socio-economic effects. As mentioned above, thanks to CAP supporting policies, renewable energy production positively affects the agricultural sector.

As mentioned no direct support is provided for dedicated biomass, but particularly through rural development plans, specific measures are provided to support energy production from renewable sources. Nevertheless, the possible impacts, synergies and trade-offs of large-scale bioenergy production and use should be studied and analyzed by connecting all aspects of agronomy with ecology, environment, economics and societal change.

The renewable energy system also plays an important role in irrigation management and more specifically in water pumping. This is considered in the ILUC framework, where the analysis of the water footprint for bioenergy production is considered as a crucial issue, adding to the agriculture for energy framework the strategic role of water.

Water resources are among the most valuable resources of the natural environment, and their sustainable and integrated management is at the basis of European water policy. With the adoption of the Water Framework Directive (2000/60/EC), EU water policy has undergone a process of restructuring. The WFD is supplemented by international agreements and legislation relating to water quantity, quality and pollution.

In particular, the WFD introduces a holistic approach for the management and protection of surface waters and groundwater based on river basins, in order to promote sustainable water use, protect the aquatic environment, and improve the status of aquatic ecosystems.

However, many obstacles still hamper a complete protection of Europe's water resources. Pollution from point and diffuse sources, over–abstraction, and alterations of rivers and lakes threaten efforts to achieve a good status for European waters.

Therefore, we need an integrated approach to the future of water resources in Europe. Under this vision, in 2012 the EU Commission launched the Blueprint to Safeguard Europe's Water Resources (COM(2012)0673), which aims at ensuring the availability of a sufficient level of quality water for all legitimate uses by better implementing current EU water policy, integrating water policy objectives into other policy areas, and filling gaps in the current framework.

References

Bruinsma, J. 2009. The resource outlook to 2050, paper presented at the expert meeting on How to feed the world in 2050, 24–26 June 2009.

Chel, K. 2011. Renewable energy for sustainable agriculture. Agronomy for Sustainable Development, Springer Verlag/EDP Sciences/INRA 31(1): 91–118.

Commission of the European Communities (CEC). 2000. Directive 2000/60/EC of the European Parliament and of the Council of 23 October 2000 establishing a framework for Community action in the field of water policy. Official Journal of the European Communities L327. Office for Official Publications of the European Union, Luxemburg, LU.

Commission of the European Communities (CEC). 2007. Communication from the Commission to the European Parliament and the Council. Addressing the challenge of water scarcity and droughts in the European Union. COM (2007) 414 final. Brussels, BE.

De Fraiture, C. et al. 2007. Looking ahead to 2050: scenarios of alternative investment approaches. *In*: Molden, D. (ed.). Water for Food, Water for Live.

Dworak, T., Berglund, M., Laaser, C., Strosser, P., Roussard, J., Grandmougin, B., Kossida, M., Kyriazopoulou, I., Berbel, J., Kolberg, S., Rodríguez-Díaz, J.A. and Montesinos, P. 2007.

EU Water Saving Potentials. Final report. Ecologic Institute for International and European Environmental Policy. Berlin, DE.

Falkenmark, M. 2003. Freshwater as shared between society and ecosystems: from divided approaches to integrated challenges. *Philosophical Transactions of the Royal Society B: Biological Sciences* 358(1440): 2037–2049.

FAO. 2014. The Water Energy Food Nexus—A New Approach in Support of Food Security and Sustainable Agriculture, Rome 2014.

Hoff, H. 2011. Understanding the Nexus. Background Paper for the Bonn2011 Conference: The Water, Energy and Food Security Nexus. Stockholm Environment Institute, Stockholm.

FAO. 1996. Rome Declaration on World Food Security and World Food Summit Plan of Action. World Food Summit 13–17 November 1996. Rome.

Global Water Partnership (GWP). 2012. Increasing water security—a development imperative. Perspective paper. Stockholm, Sweden: GWP.

Lele, U., Karsenty, A., Benson, C., Fétiveau, J., Agarwal, M. and Goswami, S. 2013. Changing roles of forests and their cross-sectorial linkages in the course of economic development: background paper 2. New York, NY, USA: United Nations Forum for Forests (UNFF).

Internet

http://www.wrc.org.za/News/Pages/Internationalandlocalwaterspecialistsconvenetotackle freshwatergovernanceissues.aspx.
https://ec.europa.eu/jrc/en/event/conference/launch-e3p.
http://ec.europa.eu/agriculture/bioenergy/index_en.htm.

Chapter-6

Water-Energy-Food Nexus in the Arab Region

Amani Alfarra*

INTRODUCTION

This chapter describes the interlinkages between water scarcity, food security and access to renewable energy in the Arab Region focusing on Jordan as a case study. It explores the challenges and opportunities of water, energy and food in the region. It therefore reviews different existing frameworks and approaches that deal with the interactions between different sectors. Using FAO's concept note on the Water-Energy-Food Nexus as a basis, the chapter takes stock of the situation in the Arab Region—the state of resources, current policies, political barriers and competing interests, particularly when it comes to ensuring stable water supply and economic growth. This will be explored in more depth, focusing on Jordan as the case study.

Water, Energy and Food Security in the Context of International Development

One of the major concerns that currently dominates the political and socio-economic debate in the Arab Region[1] is how to ensure continued water, energy and food supply (UNDP, 2013b). The region's natural resources, especially water, energy and land, are already experiencing significant stress due to the limited natural availability, growing levels of consumption, commercial and industrial demand, rapid population growth and urbanization (Elasha, 2010b; Khoday, 2011; UNDP, 2011, 2013b). The region is ranked among the most vulnerable to climate change in the world (Climate

* Water Resource Officer, CBL Division—Room B723 bis FAO of the U.N. Viale delle Terme di Caracalla, 00153 Rome, Italy. www.fao

[1] Defined for the purposes of this paper to include the 21 Arab countries (Algeria, Libya, Mauritania, Morocco, Tunisia, Bahrain, Kuwait, Somalia, Djibouti, Egypt, Iraq, Jordan, Lebanon, Oman, Saudi Arabia, United Arab Emirates, Palestine, Yemen, Syria, Comoros, Qatar).

Change Index, 2014). The 2011 uprisings and subsequent humanitarian crises in Syria and Libya have added further pressure on an already strained resources base. Nevertheless, planning processes, policy development and governance in the Arab countries are centralized with limited integration between different sectors and public (Mirkin, 2010; UN-HABITAT, 2012; UNDP, 2011, 2013b; Verner, 2012). Before going into further detail, this section provides definitions and an overview of the role of food, water and energy in supporting human livelihoods and economic development.

A widely accepted definition of **food security** was presented at the 1996 World Food Summit as "physical and economic access by all people at all times to sufficient, safe and nutritious food to maintain a healthy and active life" (FAO, 1996). In the 2009 Declaration of the World Summit, this was amended as: "Food security exists when all people, at all times, have physical, social and economic access to sufficient, safe and nutritious food, which meets their dietary needs and food preferences for an active and healthy life" (FAO, 2009). The reference to ensuring food security "at all times" is particularly significant as it implies that uninterrupted access to food has to be ensured today and in the future. This embeds the concept of sustainability and resilience in light of droughts and other extreme weather events, political instability and economic crisis.

Food insecurity exists when people do not have adequate physical, social or economic access to food as defined above (FAO, 2003). Food security is a complex condition. Its four dimensions—availability, access, utilization and stability—are better understood when presented through a suite of indicators. Improving food security requires a range of food and nutrition-enhancing interventions in the agriculture sector, but also in health, hygiene, water supply, energy and education, particularly targeting women.

Water resources play a crucial role in the web of food, energy, climate, economic growth, and human security challenges that the world economy faces over the next two decades (Beck and Walker, 2013). A common definition for water security is "access to safe drinking water and sanitation; and to water for other human and ecosystem uses" (Hoff, 2011). Wise management of water is going to be crucial for food security and sustainable agriculture (FAO, 2014), water as a constraint or enabler for economic growth and development in the short, medium and long term future. Decisions on how to conserve, manage and use water need to be made on the basis of commonly accepted and scientifically robust definitions and water accounting methods. This involves the assessment of (efficient) use of water at field, irrigation scheme and river catchment scales; the consideration of additional dimensions of productivity; and macro-economic assessments of the water-related agricultural economy contribution to GDP and global trade.

Demographic growth, economic development and climate change are putting unprecedented pressure on renewable, but finite, water resources in the Arab region (UNDP, 2013b). Water scarcity has been defined in the FAO report "Coping with water scarcity—an action framework for agriculture and food security" as:

The gap between available supply and expressed demand of freshwater in a specified domain, under prevailing institutional arrangements (including both resource 'pricing' and retail charging arrangements) and infrastructural conditions. Scarcity is signaled by unsatisfied demand, tensions between users, competition for water, over-extraction of groundwater and insufficient flows to the natural environment (FAO, 2012a).

In the Arab region, water use has been growing at more than twice the rate of population increase. Climate change adaptation will be played out to a large extent in the agricultural sector where most of the water is used. Competition with other users will grow, transferring water resources away from agriculture and leaving less water for food production (UNDP, 2009, 2013b; Verner, 2012).

At the same time, **energy** is central to addressing issues of water scarcity, food insecurity and climate change mitigation and adaptation. In the agriculture sectors, energy is required to produce, transport and distribute food as well as to extract, pump, lift, collect, transport and treat water. The water sector, too, is becoming more energy-intensive, using energy for desalination and pumping deeper groundwater, for instance. This may not seem like much of a problem in a region well-endowed with oil and gas reserves. Nevertheless, there is significant scope for improvements in terms of (a) energy use efficiency in the water and agriculture sector; (b) development of alternative sources of energy that are less carbon-intensive and can provide access to energy locally. A Nexus perspective that takes into account the interlinkages of the energy sector with water, food and climate can help to identify trade-offs and highlight synergies, so that energy security for all—as defined by the International Energy Agency (IEA) as "the uninterrupted availability *of energy sources at an affordable price"*—can be achieved.

Understanding different water, energy and food systems as well as the influence of trade, investment and climate policies is a first step towards a more coherent and forward-looking governance approach—beyond sectoral, policy and disciplinary silos. Nevertheless, there still is a relatively limited understanding of how to tackle these complex relationships when designing policy and taking action (Beddington, 2009; Hoff, 2011; WEF, 2011).

FAO's Conceptual Framework on Water-Energy-Food Nexus

The Water-Energy-Food Nexus emerged as a broad conceptual approach, inspiring a range of research initiatives, assessments and projects that all provide their own spin on the Nexus concept. Interpretations of what constitutes the Nexus are plentiful as are the analytical and methodological frameworks.

The World Economic Forum has been among the first organizations to identify the water-energy-food nexus as a key development challenge, calling for a better understanding of the inter-linkages between water, energy and food at the 2008 Annual Meeting in Davos (WEF, 2011a). The World Economic Forum Water Initiative subsequently published a book, exploring the topic of water security in relation to energy and food systems as well as climate, economic growth and human security and the Water Resources Group launched a nexus initiative with water security as a practical entry point (WEF, 2011). In line with this, the Bonn Nexus Conference in 2011 also emphasized the centrality of water resources (Hoff, 2011).

However, the focus on water bears the risk of undermining the original intention of developing an explicitly cross-sectoral perspective that could supersede traditional sectoral approaches. Isolated sectoral investments risk prioritizing the goals of one specific sector—in this case, water—over others.

Following the Bonn Conference, the Nexus concept has become more widely accepted. Several frameworks and methodologies have looked at the inter-linkages between water, energy and food (Mohtar and Daher, 2012; IISD, ADB, UN-ESCAP) (Mohtar and Daher, 2012; Bizikova et al., 2013; ADB, 2013; UN-ESCAP, 2013), but also land and soil (European Report on Development, 2012; Hoff et al., 2013), minerals (Transatlantic Academy), and ecosystems (ICIMOD, 2012; UNECE, n.d.). Although this has helped to broaden the focus, the approaches differ greatly in their scope, objectives and understanding of drivers as the following two examples illustrate.

The Qatar Environment and Energy Research Institute (QEERI) developed a conceptual framework and tool that addresses the Qatari context of water scarcity, arid lands and a high dependency on food imports (Mohtair and Daher, 2012). Qatar's food security is the central focus, imputing the relative importance of financial sustainability, water availability and CO_2 emissions as complementary development objectives (QEERI, 2012). The Nexus inter-linkages are described in terms of the availability of resources (in terms of water and energy by source and land requirements) needed to produce and import food. It does not consider issues of quality, access or stability of supply nor does it look at resource uses other than for food production.

The European Report on Development (ERD) (European Report on Development, 2012) focuses on the management of water, energy and land resources across sectors subject to increasing relative scarcity and prices. The estimated impacts of climate change, global population and economic growth are also considered. The entry point for analysis is the integrated management of water, land and energy resources. The report examines the constraints on resources and how they can be better managed to support inclusive and sustainable growth. It examines, for example, the implications of land management policies on different natural resources. It does not, however, articulate systemic—"nexus-wide"—ripple effects of such policies on baseline conditions, for which they had been initially designed.

Departing from a static, water-centric interpretation of the Nexus, FAO has developed a conceptual approach to make sense of, and manage the complex and interlinked uses of water, energy and food (Fig. 6.1). The approach distinguishes between the resource base and the different goals and interests that are to be achieved with the same, but limited resources. It is about understanding and managing these different resource user goals and interests, while maintaining the integrity of ecosystems. For that, FAO has identified three areas of work through which it can contribute, namely by providing evidence, by developing scenarios, and by designing and appraising response options.

Evidence is needed to inform and substantiate discussions on future scenarios and responses. Similarly, scenarios provide a way to explore the potential impacts of different responses. None of this can be done in isolation, but only in dialogue with stakeholders.

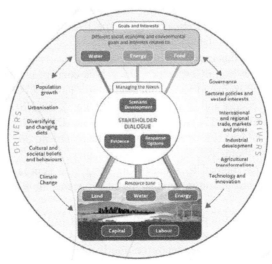

Figure 6.1 A conceptual diagram of the FAO approach to the Water-Energy-Food Nexus.

(1) **Evidence:** Reliable, relevant and timely data and analysis of key issues is needed. FAO works to provide data, tools and methods—ranging from Earth Observations to statistics and indicator-based assessment tools—in order to analyze the current state and use of natural resources, trends and variability.

(2) **Scenario Development:** Scenarios can be useful to explore strategic questions, to review policies and investment decisions, and to create common ground and improved understanding of the interrelations between water, energy and food.

(3) **Response Options:** Moving from understanding to action, FAO can play an important part in planning, implementing and evaluating response options, such as policies, investment decisions, regulations and incentives (including subsidies, promotion of appropriate business models, specific financing facilities and other support mechanisms), capacity development and training, and technical interventions.

Many valid sectoral approaches exist for the development of sustainable food and agriculture and there continues to be a great need to find solutions within sectors, for example, by improving water use efficiency. At the same time, we need a shared vision and effective mechanisms to deal with cross-sectoral issues.

Arab Region: "Water Scarcity and Food Security" Nexus

Over the last four decades, the Arab world has experienced a development boom, with rapid population growth (UNDP, 2011). To meet the parallel rise in demand for food, many Arab countries have prioritized food security and socio-economic development through policies to expand agricultural land and irrigated cultivation. However, they have not adequately considered the limited availability of water resources and the need for conservation and demand management (AFED, 2010;

El-Naser, 2013; UNDP, 2013b). Water scarcity has become a critical constraint to agriculture (AFED, 2010).

Demographic trends

- Since 1970, the population of the Arab world has nearly tripled, climbing from 128 million to 357 million in 2010, with 56% living in cities. According to the United Nations it is expected to reach 646 million inhabitants by 2050, while urbanization will continue at an accelerated pace and reach 68% in 2050 as illustrated in Fig. 6.2 (UN-HABITAT, 2012; UNDESA, 2011).

Rapidly growing population exerts pressure on natural resources, agricultural land, contributes to environmental degradation, raises demand for food and shelter which encourages the conversion of forest land for agricultural and residential uses (Ahmad et al., 2005). Therefore, in order to feed this larger, and more urban population, it is estimated that water withdrawal and food production will be doubling in the Arab region (FAO, 2006).

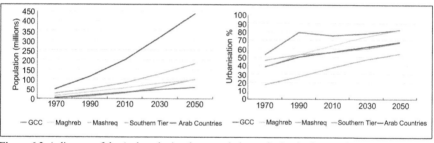

Figure 6.2 A diagram of the Arab region's urban population and urbanization trends (1970–2050) (UN-HABITAT, 2012; UNDESA, 2011).

Color version at the end of the book

Environment, climate and climate change

The Arab region faces complex and serious environmental challenges, which mainly include water scarcity; very low and variable precipitation; limited arable land; excessive exposure to extreme weather events, including droughts, desertification and other climate change impacts (UNDP, 2011). This demanding environment, combined with fast population growth makes the Arab region among the world's most vulnerable regions to face food insecurity (Abdel-Dayem and McDonnell, 2012; Breisinger et al., 2012).

Almost 90% of the Arab region's lands are classified as hyper-arid, arid and semi-arid land zones. The average annual temperatures, as well as maximum and minimum temperatures, also vary from freezing to over 50 degrees Celsius, depending on the season and location (Abahussain et al., 2002). Among the most influential climatic factors is rainfall (UNDP, 2013b). There is a severe spatial rainfall distribution over the Arab region; around 52% of the region's area receives an average annual rainfall of less than 100 mm, while 22% receives 100–200 mm, 15% receives 200–500 mm, and 11% of the region's area receives more than 500 mm of rainfall on an annual basis (Abahussain et al., 2002). Future rainfall scenarios demonstrate that most of the Arab region will receive less rainfall averages by 15% by 2088 and 30% in 2099 (UNEP, 2013).

A recent study conducted by Arab Center for the Studies of Arid Zones and Dry Lands (2010), shows that almost 80% of the region's area was experiencing 0–4 years of drought in the last 10 years (2000–2010) (Erian, 2010). Areas that receive an annual amount of rainfall range between 120/150–400 mm are considered vulnerable areas to drought (Erian et al., 2006; ACSAD, 2008; Droogers et al., 2012). The Arab world's dry lands support the livelihoods of 60% of the total population (Erian, 2010).

Most recent assessments have concluded that arid and semi-arid regions are highly vulnerable to climate change (IPCC, 2007, 2013; Verner, 2012). There is an increasing evidence that climate change will have severe negative impacts on the water availability, decrease agriculture productivity by 10 to 40%, and hinder the economic and social development of Arab countries (Mirkin, 2010; Verner, 2012; Verner and Breisinger, 2013; Verner et al., 2013).

Water scarcity

The Arab region holds 1% of the world's water resources (UNDP, 2013b). Water scarcity is alarming in the region, since it has the lowest freshwater resource endowment in the world (Mirkin, 2010; Verner, 2012). The region receives an estimated 2,282 billion cubic meters (m^3) of rainwater each year compared to an estimated 35 billion m^3/year of groundwater and 205 billion m^3/year of surface water of which only 43% of which originates within the Arab countries (Elasha, 2010b).

Arab countries are considered water-scarce, with consumption of water significantly exceeding total renewable supplies. The region has less than 500 cubic meters of renewable water resources available per person annually (FAO, 2015). About 66% of Africa is arid or semi arid, and more than 300 million people in sub-Saharan Africa live on less than 1,000 cubic meters of water resources each (UNESCO, 2012).

With rapid population growth, increasing per capita use, and fast urbanization, the per capita availability of water is likely to be reduced in the Arab countries by about 50% by the year 2025, and demand is projected to increase further by 60%, by 2045 (Gober, 2010; UNEP, 2013; Verner, 2012).

This scarcity and declining of water quality in many Arab world countries will increase the competition over water within and between sectors (UNDP, 2013b). This competition will transfer water out of agriculture and leave less water for food (UNDP, 2009, 2013b; Verner, 2012), increase the access inequity, and escalate tension and conflicts at local, national and transnational levels (Al-Awar et al., 2010; Elasha, 2010a; UNDP, 2013b).

Water scarcity and food security linkages

Water and food are closely entwined in the Arab world (Larson, 2013). From 2000 to 2010, water use within the region has been growing at more than twice the rate of population increase. At the same time, there is an increase in the number of countries which are reaching the limit at which reliable water services can be delivered (AWC, 2011). Basically, demographic growth, economic development, and climate change are putting unprecedented pressure on renewable, but finite, water resources (UNDP, 2013). This scarcity further aggravates pressures on the agricultural sector as it competes with other sectors, transferring water out of agriculture and leaving less water for food production (UNDP, 2009, 2013b; Verner, 2012). Currently, water and food supply in the Arab region are highly unstable, driven by energy prices, poor harvests, biofuels, rising demand from a growing population, climate change, economic crises, and political tensions (Abdel-Dayem and McDonnell, 2012; Breisinger et al., 2012). Molden (2007) developed a conceptual framework (Fig. 6.3) that illustrates the interlinkages between these dynamics aiming at reducing poverty and hunger and ensuring environmental suitability (Molden, 2007).

Agriculture is a major as well as sensitive sector of the region's economy (UNDP, 2011, 2013b). The current total cultivated area makes up about 5% of the total global cultivated area, and represents about 5% of the total land area in the Arab region (FAO, 2006). Although up to 90% of the region's water is extracted for agriculture as illustrated in Fig. 6.3, most countries in the region import more than 50% of their caloric intake (Abdel-Dayem and McDonnell, 2012). The agricultural performance indicators show that irrigation management is weak, characterized by deteriorating infrastructure; centralized administration; large irrigation bureaucracy; low irrigation service fees and limited participation of water users in maintenance tasks (UNDP, 2013b).

Agricultural production in Arab countries is projected to grow by more than 60% between 2001/2003 and 2030 and more than double by 2050 (FAO, 2006). Exactly how much water will be needed to meet projected food demand is still debatable, however, studies suggest that at least 20% more irrigation water will be needed by 2025 (Abdel-Dayem and McDonnell, 2012; FAO, 2006). This will represent a great challenge to water-scarce countries with limited surface water resources and non-renewable groundwater aquifers depleted at unsustainable rates, such as Saudi Arabia and Jordan (Kfouri, 2013). In addition to overexploitation,

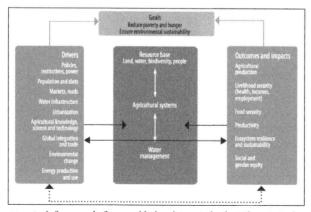

Figure 6.3 A conceptual framework for considering integrated adaptation strategies for addressing agriculture, water, food security, rural livelihoods, and environment issues (Molden, 2007).

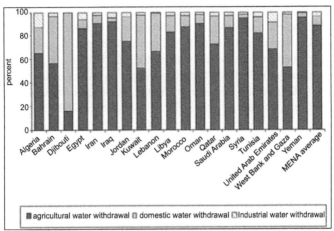

Figure 6.4 A diagram of the water withdrawal per sector in the Arab countries (Abdel-Dayem and McDonnell, 2012).

pollution from agricultural, industrial and domestic activities threatens the Arab region's groundwater and surface water resources. As water quality deteriorates, water usability diminishes, reducing water supplies, intensifying water scarcity, increasing health risks and damaging the environment, including fragile ecosystems.

Water demand competition is also growing among other sectors, including those of expanding urban centers, industry, energy production, and aquatic systems (UNDP, 2013b). The aforementioned water uses leave nations with serious water security challenges and may raise conflicts between stakeholders. Balancing water security and food security within the region is a big challenge, particularly in countries with limited water resources and fast-growing populations. This emphasizes the emerging need to develop mitigation and adaptation policies (Verner, 2012; Verner and Breisinger, 2013).

In the absence of other economic opportunities in rural areas, poverty reduction is closely linked to water development for irrigated agriculture (FAO, 2007). Water and food shows strong interconnectedness (Bazilian et al., 2011), as they have become two sides for one coin. There is a policy and institutional dimension to this part of the nexus that needs to be properly communicated through a common global, regional, national and local agenda in order to relax any stresses water and food systems are encountering due to current practices, and to meet the increasing demands without exerting more pressure on the nexus (Mohtar and Daher, 2012; WEF, 2011). Wise management of water is going to be crucial for food security and sustainable agriculture (FAO, 2007), water as a constraint or enabler for economic growth and development in the short, medium and long term future. Water is gaining traction as an element of risk in investment decisions.

Energy and food security linkages

The energy sector in the Arab region has been and will continue to play a critical role in the region's socio-economic development (Bachellerie, 2012). The Arab Monetary

Fund's Report (2010) shows that the oil and gas sector makes up to about 38% of the total Arab gross domestic product (GDP) (AMF, 2010). Moreover, the region relies heavily on oil and gas to meet more than 97% of domestic energy demand (AFED, 2013). There are, however, significant differences in energy endowment within the region and many of the benefits of the region's oil and gas wealth remain unequally distributed across the Arab world. Accordingly, the six GCC member states along with Algeria, Iraq, and Libya account for nearly 98% of the Arab world's total oil reserves, and 95% of its total production, while others are less well endowed. Though intra-regional trade and investment flows, intra-Arab aid, and the long-term flow of labor remittances by the Arab expatriate workforce brings some benefits to the region.

Within the region, electrification rates vary—between 100% in some countries, such as Kuwait, and very low rates like in Mauritania or Yemen. In terms of cooking and heating, almost a fifth of the Arab population relies on non-commercial fuels like wood, dung, and agricultural residues, particularly in Sudan, Yemen, and Somalia but also in Algeria, Egypt, Morocco, and Syria. Moreover, access to modern energy and energy prices are a key factor for food insecurity in many of the Arab countries—excluding GCC countries (UNDP, 2013b). For instance, in Yemen it has been estimated that 70 to 80% of Yemen's cereal requirements alone are imported. Recently food security worsened due to soaring domestic food prices as a result of high international food prices and fluctuation in energy prices. Yemen's population has been affected by the energy prices due to its direct impact on agricultural practices and productivity. For example, lack of income or high fuel price, has widely hindered the widespread use of better irrigation techniques (diesel-powered water pumps), reliance on rainfall (38.8% of household) for cultivation of crops was significantly associated with increased food insecurity at the rural household level (El-Katiri and Fattouh, 2011).

Currently, the Arab region contributes about 5% of global GHG emissions (AFED, 2011; Khatib, 2010). Although this seems relatively low, the GHG emissions of the region are expected to reach 9% of the world's total by 2035 (Khatib, 2010). The World Bank (2007) identified that 85% of GHG emissions in the Arab region come from energy production, transformation, and use in the Gulf Cooperation Council (GCC) countries (WB, 2007). As such, the energy sector plays a key role in the mitigation of climate change impacts (Habib-Mintz, 2009). The Arab region cannot afford inaction at global, regional, or national scale, particularly because the region is considered so highly vulnerable to projected climate change impacts (AFED, 2009; UNEP, 2013).

In light of this, it is important to remember that the Arab region has a huge potential for renewable energy generation, mostly wind and solar (AFED, 2011). Yet, its renewable energy power sector remains underdeveloped. In 2009, the share of renewable energy in the total energy mix in the region was around 4% (OECD, 2013). A gradual shift to renewable energy sources can bring unique economic opportunities and carbon reduction to the Arab region (AFED, 2011; UN, 2009).

According to the International Renewable Energy Agency (IRENA) recent publication "Renewable Energy in the Water, Energy and Food Nexus", indicated that renewable energy technologies could address some of the trade-offs between water, energy and food, bringing substantial benefits in all three sectors. They can allay competition by providing energy services using less resource-intensive processes and technologies, compared to conventional energy technologies. The distributed nature of many renewable energy technologies also means that they can offer integrated solutions for expanding access to sustainable energy while simultaneously enhancing security of supply across the three sectors (IRENA, 2015). Figure 6.5 below illustrates the entry points for renewable energy into conventional energy supply systems.

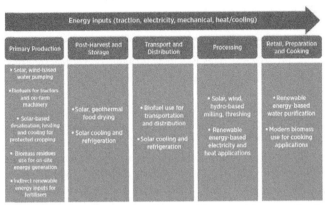

Figure 6.5 Illustration of different entry points for renewable energy into conventional energy supply systems (FAO, 2011; IRENA, 2015).

Innovative Policies, Practices and Technology

The water crisis is a crisis of governance (UNDP, 2004). Water governance is defined by the United Nations as *"political, social, economic and administrative systems that are in place, and which directly or indirectly affect the use, development and management of water resources and the delivery of water service delivery at different levels of society"* (UNDP, 2003). Importantly, the water sector in the Arab world is a part of broader social, political and economic developments and is thus also affected by decisions outside of the water sector (UNDP, 2013b).

Ineffective governance institutions, lack of transparence and corruption has prevented governments from providing adequate protection against increased water scarcity in most Arab countries (El-Naser, 2013). Similarly, The United Nations Development Programme Report (2013) on "Water Governance in the Arab world" concluded with the following guiding principles and recommendations can help realize effective water governance in the Arab world: decentralization, reorienting policies, instituting reform, addressing inadequate and weakly enforced legislation, empowerment, sustainability, address water-related challenges and nexuses (UNDP, 2013b). Most Arab countries have developed the institutional and legislative framework for good water governance but still lack legislative instruments to support its implementation (El-Naser, 2013).

Agricultural practices and technologies can help to cope with water scarcity

Water security is inseparable from other critical issues, such as food and energy security. In severe water-scarce regions, efforts and policies would have to focus on getting "more crops per drop", taking into account there are no new water resources for irrigation; therefore any expansion in agriculture must come from water savings (Abdel-Dayem and McDonnell, 2012; Gregory et al., 2005; Shetty, 2006). One solution adopted by mostly the oil-exporting countries of the region is **to desalinate water resources**. The Arab region holds about 60% of the total global desalination capacity. Given the high costs and energy requirements, this option is still limited to the major oil-exporting countries (Abdel-Dayem and McDonnell, 2012).

Likewise, some literature advocated the following actions in **agricultural practices** in response to water scarcity (Abdel-Dayem and McDonnell, 2012; AFED, 2010):

1. Increase the supply (through wastewater reuse, rainfall harvesting, storage, etc.);
2. Increase efficiency and productivity of water use;
3. Reduce demand (e.g., cropping allocations);
4. Enhance capacity of especially rained agriculture to cope with climate variability and change;
5. Recognize the importance of facilitation and conflict resolution techniques in discussions among stakeholders, and develop capacities to apply these.

Food trade can relieve water stress

Trading in agricultural products also means trading of the water embedded in growing these products, which is known as virtual water (AFED, 2010). Therefore, water should be viewed and recognized as a global resource, and the actual amount of water saved by countries for food production should be viewed as a reduction to the global water bill (Daher, 2012). A major example of virtual water import is Jordan. Jordan imports close to 90% of its food (Forbes, 2008). Importing five to seven billion cubic meter of water in virtual form per year is in sharp contrast with the 1 billion m^3 of water Jordan withdraws annually from its domestic water sources (Haddadin, 2003). Another example is Egypt, the largest wheat importer in the world, importing about 45% and 20% of their need from USA and Australia respectively. As a result, Egypt saves its scarce blue water resources. Greater integration with global markets will become essential (Daher, 2012). Reforms of trade policy may provide an important "win-win" policy for water conservation (Richards, 2001).

Rethinking subsidies can reduce trade-offs between sectors

During the last three decades, government's policies in the Arab region ignored water scarcity and strongly encouraged agriculture development with subsidies on water and fuel and low interest rates on loans for digging new water wells or securing equipment (UNDP, 2011, 2013b). This led to unsustainable water abstraction with perverse implications for economic development and social welfare of some Arab countries (UNDP, 2011). One of the easiest ways to improve the efficiency of irrigation is to stop subsidizing the prices of water-intensive crops (UNDP, 2013b). Additionally, establishing effective social safety nets, measures to tackle price volatility including appropriate use of food reserves, significant investments in modernization of irrigation and drainage systems, are steps to prevent future food crises (UNDP, 2013b).

Political and other Barriers

The Arab countries' demand for greater political inclusion and improved socio-economic security has been at the core of the recent instability. This has triggered a series of policy changes to improve living conditions in general, and food security in particular (Kfouri, 2013; UNDP, 2013a). Without appropriate adaptation responses, future projections for the Arab region indicate a severing of living conditions, particularly for the least resilient population (Verner, 2012). Likewise, Verner and Breisinger (2013) argued that over the next 30 to 40 years climate change is likely to lead to a cumulative reduction in household incomes ranging from close to US$2 billion in Syria and Tunisia and up to US$9 billion in Yemen (Verner and Breisinger, 2013).

Water management in the Arab world is centralized (UNDP, 2013b). Richard (2001) argued that decentralization and devolution would be essential for future strategies for coping with water scarcity. Ultimately, there is simply no other choice. The inefficiencies of large, top-down, supply-enhancing approaches cannot continue to deliver the same volumes of water as in the past (Richards, 2001).

Arab countries must also enhance coordination and investment in R&D in water and renewable energy technologies, most of which are currently imported. Acquiring and localizing these technologies will make them more reliable, increase their added value to Arab countries' economies and reduce their cost and environmental impacts (UNDP, 2013b).

Water supply, economic growth, and border tensions

The relationship between water supply and economic growth is still controversial Howe (1976) argued that water is a bottleneck to the continued growth of economic activity; the provision of more water, better-controlled water, or better quality water constitutes a necessary condition for economic growth. There is no guarantee, though that such a provision will be sufficient to initiate further growth (Howe, 1976). On the contrary, Barbier (2004) suggests that current rates of fresh water utilization in the vast majority of countries are not yet constraining economic growth. However, countries that are 'water stressed' may find it difficult to generate additional economic growth through more water use (Barbier, 2004). Others have argued the strong relation and indicate that the cause of the global water crisis is largely the result of population growth and economic development rather than global climate change (Vörösmarty et al. 2000).

Homer-Dixon (1994) developed a conceptual framework that illustrates the relationships between the sources of environmental scarcity and social effects as illustrated in Fig. 6.6, and argued that a decrease in quality and quantity of renewable resources accompanied by population growth and unequal access to resources would increase environmental scarcity, limit economic productivity, weaken states and result in tensions and conflicts (Homer-Dixon, 1994). For example, one of the underlying causes of the conflict in the Darfur region of Sudan was the water shortages, drought, and desertification. It is important to recognize that food security and ecological sustainability are closely intertwined and that both are critical to human security (Gonzalez, 2004; Idriss et al., 2012).

Scarcer water supply in the Arab region will contribute to economic, social and political instability (UNDP, 2013b). However, water-related problems—when combined with climate change, poverty, social tensions, environmental degradation, ineffectual leadership, and weak political institutions—contribute to economic and social disruptions that can result in the state's failure. The recent Arab uprising could be a good example (El-Naser, 2013). Additionally, the lack of adequate water will be a destabilizing factor in some Arab countries because they lack financial resources or technical ability to solve their internal water problems, for example (e.g., Yemen, Somalia, and Djibouti) (AWC, 2004). In addition, some countries are further stressed by a heavy dependency on river water controlled by upstream nations (e.g., Iraq, Syria, and Egypt) with unresolved water-sharing issues (UN-ESCWA and BGR, 2013). Wealthier countries probably will experience increasing water-related social disruptions but are capable of addressing water problems without the risk of state failure (e.g., Gulf region) (Erian, 2010). Renewable energy technologies offer an opportunity to decouple water production from fossil fuel supply, and to cater to the heat or electricity needs of desalination plants. Recognizing this opportunity, Saudi Arabia announced King Abdullah's Initiative for Solar Water Desalination in 2010, which aims to enhance the country's water security and contribute to the national economy by developing low-cost solar-based desalination technology. Although desalination based on renewable energy still may be relatively expensive, decreasing renewable energy costs, technology advances and increasing scale of deployment, might soon make it a cost-effective and sustainable solution.

Additionally, competition for water within the Arab region and across its borders expected to grow, carrying the risk of conflict (FAO, 2008). As 57% of the surface

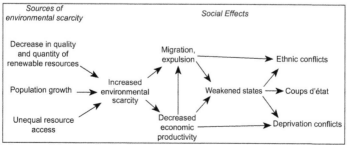

Figure 6.6 Conceptual framework of environmental scarcity sources and consequences (Homer-Dixon, 1994).

water originates from out of the region, and many of the major water resources in the region are shared between countries lying both within and outside the region (Mirkin, 2010; Verner, 2012). The most significant basins are those of the Jordan, Nile, and Euphrates/Tigris, all of which are subject to contentious riparian issues (UN-ESCWA and BGR, 2013).

The water conflicts issue is still debatable, Tignino (2010) argued that water scarcity, accelerated by population growth and climate change, affects water availability, competition between multiple uses and may threaten peace and security (Tignino, 2010). While some predict such conflicts, others indicate that no conflicts have taken place since 4,500 years (Philips et al., 2006). On the contrary, increased competition for water could become the catalyst for more intense co-operation in the future (Starr, 1993). However, other researchers argued that interactions between riparian states over shared water resources, include forms of both conflict and cooperation between states (Philips et al., 2006; UNDP, 2006; UNESCO, 2003; Zeitoun and Mirumachi, 2008). Ensuring access to water, along with the protection of water resources, contributes to preventing conflicts and to the transition to peace in post-conflict states (Bruch et al., 2008).

Addressing water challenges now can also help strengthen the economy and resilience by managing the risk of potential crises that could result from inaction: such as unplanned migration, economic collapse, or regional conflict (UNDP, 2013b).

Jordan—Case Study

The Jordan valley is one of the most heavily populated, food insecure and water-scarce regions in the world, placing the region's energy and freshwater resources under severe stress. The water issue in Jordan has a complex history combined with ongoing challenges due to the influx of Syrian refugees and by regional disturbance.

As a result, the once mighty Jordan River has been reduced to just 2% of its original flow and underground aquifers are being pumped at unsustainable rates. Compounding these issues further, current climate models predict that the Jordan River Basin will become hotter and drier, increasing concerns over water supply, energy production, food security, and international relations. For example, the 2014 Jordan's Third National Communication Report on Climate Change indicated that the rainfall amounts and climatic conditions of the country do not support good rainfed agriculture, except for few areas in the northern and western highlands. The rainfed agricultural zone lies in areas where rainfall exceeds 250 milimeters although significant production of cereals does occur in some areas where rainfall is between 200 and 250 millimeters (UNFCCC, 2014).

In the past half-century, regional instability has been one of the major obstacles to achieving sustainable development in the region. However, more recent experiences demonstrate that cooperation over food, water and energy issues in the region is in fact attainable and may even help drive greater regional stability. Regional policymakers and the global community alike have an obligation to work towards sustainable solutions for food, water and energy within the Jordan River Basin, mainly to support Jordan in it regional role by hosting huge numbers of refugees from neighboring countries due to regional turmoil (Meisen and Tatum, 2011).

Syrian Refugee and the international response to Syria crisis

The security situation of people in Syria Lebanon, Jordan, Iraq, Turkey and Egypt is critical and deteriorating daily. In Syria, food supplies are limited, difficult to access and costly. Fields and farming assets have been left idle or destroyed due to violence, displacement, increased production costs and a lack of basic farming supplies. The wheat harvest in 2013 fell 40% short of an average year. Syria's essential services are on the brink of collapse under the burden of continuous assault on critical water and energy infrastructure. This situation has been further aggravated by the eighth consecutive summer of drought that could escalate in a full-blown water, energy and food crisis in its own right. With over three million refugees, host communities in neighboring countries face a huge demand and intense competition for resources, such as land and water, and income opportunities, while costs for housing, food and other basic commodities soar.

As the crisis shows no signs of abating, the response is now shifting from humanitarian relief to longer-term, development-focused response. In Jordan the shift has taken place in form of the 2015 Jordan Response Plan for the Syria Crisis and the 2014–2016 National Resilience Plan. These plans encompass a great number of sectors in an attempt to mitigate the impacts of the crisis as well as to build back more resilient livelihood systems. They mark a shift in the way we deal with resources. While basic access to water, energy and food supply still has priority, the focus is now on the development of sound supply and management structures. Nexus thinking can help identify synergies as, for example, in the Zaatari Camp in Jordan, where basic water and energy resources are needed on a continuing basis by the 84,000 people currently living in the camp. Renewable energy capacities are being developed to power two planned wells to access the groundwater resources on which the camp is situated. As the crisis is seemingly protracted, this is a cross-point for humanitarian and development communities to work together more constructively to ensure that their programs are reinforcing and not conflicting or disrupting each other.

To what extent do national development strategies need to be adjusted to deal with the extra strains the Syrian crisis has put on resources, and more importantly, how the needs of the increasing number of refugees can be met in the long-term.

The Za'atari camp is the largest refugee camp in Jordan, was first established in the summer of 2012 to host 81,000 Syrians fleeing the violence in the ongoing Syrian civil war that erupted in 2011. The camp lacks a proper sanitary system and depends on primitive cesspools within the camp after passing through small ditches and lagoons between the caravans and tents and was evacuated periodically through septic tanks away from the camp polluting the surface and groundwater of the camp site, this continued to deplete natural resources and pollute ecosystems at a rate that is unsustainable and this will compromise the capacity for nations to produce food for future generations.

Water

Jordan has limited water resources; it is among the lowest in the world on a per capita basis at 147 m^3 per person per year in 2010 (MWI, 2011). Renewable water resources have fallen below 130 m³ per person per year. Current total uses exceed renewable supply. The difference (the water used that is not renewable) comes from nonrenewable and fossil groundwater extraction and the reuse of reclaimed water. Water availability per capita ranks lowest in the world and all renewable water resources of suitable quality are fully exploited. Water-scarcity in Jordan seriously affects the social and economic development of the country.

The limited water resources are exposed to pollution and increasing the pressure on available water resources. The situation is likely to exacerbate as the population with unpredicted rates based on political instability combined with climate change scenarios indicate a significant reduction in water quantity. The threat of depleting water resources that can no longer meet the increasing demand might create political instability in Jordan and wreak havoc on future generations. Due to the ongoing Syrian situation the number of Syrian refugees is expected to keep increasing up to 1.2 million Syrian refugees according to UN officials, who form 21%.

Irrigation water is heavily subsidized, with very low tariffs for surface water deliveries to the Jordan valley, and very low tariffs and little quantitative restriction of over-abstraction of groundwater in the highlands. Many studies have concluded that agricultural water use is of low economic return and that large-scale reallocation to municipal and industrial use is feasible. They cite the sector's declining contribution—3.2% t in 2012—to gross domestic product (GDP) for use of % of the country's total water supply. However, irrigation in the Jordan valley supports a large number of jobs that would be difficult and expensive to replace, uses much of the country's reclaimed wastewater that has no other current use, is trending towards higher water use efficiency, supports export-oriented value chains, and enjoys substantial political support.

Investment in wastewater treatment in Jordan has expanded over the past three decades, resulting in about 65% of the population being connected to wastewater collection and treatment systems. Currently, 27 wastewater plants serve the country. These plants processed about 105 MCM of raw wastewater in 2010 with effluent usable for irrigated agriculture of about 103 MCM.

Pressing water scarcity in Jordan increased the demands of marginal water for agriculture, of which the treated wastewater is the most prominent candidate. Meanwhile agriculture is an important economic activity in Jordan where treated wastewater could be a valuable source for irrigation in the agricultural sector. Water resource management in Jordan requires constant crisis management of dwindling quantities and deteriorating quality of supply to avoid disasters in drinking water supply and wastewater treatment.

Reallocation of water from its current 65% use in agriculture to municipal, industrial, and commercial use is evolving slowly because of the socioeconomic costs of dislocation in the Jordan valley, political resistance in the highlands, and continuing public tolerance for the current service levels of intermittent domestic water supply.

Energy

Jordan has limited domestic fossil fuel sources as most of its energy needs are imported. Jordan's native energy resources are its modest gas reserves, oil shale deposits and tar sands. It also has a small hydropower and biogas potential. In the Master Strategy of the Energy Sector in Jordan for the period (2007–2020) (the Energy Strategy), the Jordanian government set a target to obtain 1,800 MWs, or 10% of the country's energy supply, from renewable sources by 2020.

The prices of energy imports have increased with high risk in constant supplies, this situation spurred governmental action to improve energy efficiency and provide additional energy resources. Jordan imports 96% of its oil and gas, accounting for almost 20% of the GDP which makes the country completely reliable on and vulnerable to the global energy market. The current energy strategy is to transform the energy mix from one heavily reliant on oil and natural gas to one more balanced with a higher proportion such as the of energy supplied by oil shale and renewable sources. As part of this, the government opted for a gradual removal of subsidies for gasoline, diesel, fuel oil and kerosene in 2005. This was driven in part by a strategy to liberalize Jordan's energy markets as well the need to alleviate the financial burden to its economy caused by rising oil prices (MOE, 2012).

The national energy strategy seeks to increase reliance on local energy sources, from the current 4 to 25% by 2015, and up to 39% by 2020. Placing more emphasis on the utilization of renewable energies will alleviate the dependency on the traditional energy sources, especially oil which is imported from neighboring countries. This will also be paralleled with the reduction of energy produced from oil from 58% currently to 40% in 2020. On the other hand Jordan is a country with limited indigenous energy resources heavily depending on its imports of energy to meet growing demand, expected to double to a forecasted 15.08 Mtoe (million tonnes of oil equivalent) by 2020 from 7.58 Mtoe in 2007 (UNFCCC, 2014).

In March 2015, Jordan signed the agreement for 52.5 MW solar PV Shams Ma'an power plant. Shams Ma'an Power Generation company has already signed a 20-year Power Purchase Agreement (PPA) with the National Electric Power Company (NEPCO), which is Jordan's publicly owned power transmission company. In July last year, the company signed an Engineering, Procurement and Construction (EPC) contract with US-based First Solar, which will also supply the project with its advanced thin film PV modules.

The Shams Ma'an 52.5 MW project is the result of the first round of renewable energy auctions held by the Jordanian Ministry of Energy and Mineral Resources, which in total has approved 12 PV projects of various sizes and 200 MW cumulative installed capacity (the so-called round one or stage one auction has also approved two wind projects).

Jordan has a target to install 600 MW of solar PV and 1200 MW of wind capacity by 2020. Electricity production in Jordan currently depends on imported oil and natural gas. In 2011, 97% of Jordan's energy was imported and energy expenditure accounted for 20% of the country's GDP, according to the energy ministry.

Food

Due to the climatic conditions, only 1.97% of Jordan's land is arable. The Jordan Rift valley and the highlands are the nation's main agricultural areas. The extensive Arid Zone (semi-desert) and the Badia (Eastern Desert) are too dry to be cultivated.

Annual precipitation is low and largely unpredictable. Approximately 67% of Jordan's agricultural production is rain fed, leaving it vulnerable to drought. Seventy per cent of the land receives less than 100 mm of rain per year. Climate change is likely to further reduce that level. The limited water supply prevents the agricultural sector from developing. The contribution of agriculture to GDP is only 3.4% (World Bank, 2013).

Jordan's main agricultural products include: citrus fruits, tomatoes, cucumbers, olives, strawberries, stone fruits, sheep, poultry and dairy products. Current production levels are minimal, meaning it must import 97% of its food. The average diet in Jordan is based on wheat and, to a lesser extent, rice. Jordanians are undergoing a nutritional transition, progressively shifting towards a less nutritious diet, rich in fat, sugar and carbohydrates. This has had an impact on public health. Obesity is now an issue, especially among women, as is anaemia.

Generally, rainfall amounts and climatic conditions of the country do not support good rainfed agriculture, except for few areas in the northern and western highlands. The rainfed agricultural zone is in areas where rainfall exceeds 250 milimeters although significant production of cereals does occur in some areas where rainfall is between 200 and 250 millimeters.

Climate change

Article 2 of the UNFCCC refers to the dangerous human influences on climate, in terms of whether they would allow ecosystems to adapt, ensure that food production is not threatened and chart a path of sustainable economic development. Global, national and local level measures are needed to combat the adverse impacts of climate change induced damages.

According to the United Nation's Intergovernmental Panel on Climate Change, average temperatures in the Jordan River Basin are expected to increase by up to 3.1 degrees Celsius in the winter and as much as 3.7 degrees Celsius in the summer, which would cause average rainfall to decrease by 20–30% over the next 30 years. Climate scientists predict that this would result in further reductions to the Jordan River's flow, a decrease in the recharge rates of natural aquifers, cause desertification of arable land and soil erosion along the Mediterranean coast, and increase unpredictability of climatic events.

Adverse impacts of climate change will negatively affect progress toward development in a number of key areas including agriculture and food security, water resources, public health, climate-related disaster risk management and natural resources management. It anticipated that climate change will constrain the ability of developing countries to reach their poverty reduction and sustainable development objectives under the United Nations' Millennium Development Goals (MDGs). The achievement of the MDG targets will depend on effective planning for managing climate risks.

A number of constraints exist with regards to ensuring resiliency of the MDGs in the context of emerging climate change pressures.

Increasing temperatures, coupled with changing precipitation patterns, are expected to decrease surface water availability, and, acting on top of other stresses, increase water scarcity in the country.

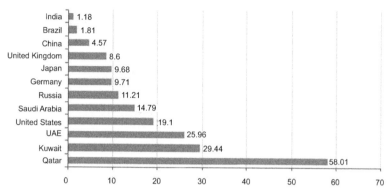

Figure 6.7 CO_2 emissions per capita (tons CO_2 per capita)—Source: Green Gulf presentation "Economic and Social Development in MENA from Renewable Energy", dated October 26th, 2010.

The MENA region meets 98% of its primary energy needs through oil and gas. (1) Oil consumption in the Middle East rose by 5.4% in 2006, faster than in any other region of the world, except China. (2) Abundant energy resources coupled with government subsidies on energy prices make energy consumption the cheapest in the region across the world. The region is one of the highest per capita carbon dioxide emitters worldwide. The MENA region's high energy intensity and sole-dependence on hydrocarbons has resulted in excessive CO_2 emissions (UNFCCC, 2012) Fig. 6.7.

There is a large potential for renewable resources in the MENA region which have remained largely untapped. This is especially true of solar power; its potential in the Middle East and North African region alone far exceeds global electricity demand.

The solar potential is huge, according to a report by the Electrical Engineering Department at King Saud University, the Middle East receives 3,000–3,500 hours of sunshine per year; with more than 5.0 kW/m^2 of solar energy per day. By applying renewable sources to energy production in the Middle East the life of fossil fuels will be expanded for future generations, decreasing carbon emissions, and encouraging socio-economic development for sustainable wealth.

Conclusion

The water security outlook of the Arab region depends on inclusive decisions made today for tackling water governance challenges, climate change, conserving water, reforming international trade, and diversification of food production. This will require policies in harmony and not at cross-purposes, planning water systems as multiple use systems, revise the legal framework and governance, and enhance the regional and global cooperation (virtual water, land acquisition). Unprecedented global cooperation is required to address the institutional, governance and financial constraints to ensure future food security for all. Addressing these interlocking issues simultaneously is inevitable to prevent famine in poor nations.

Renewable energy technologies offer opportunities to address trade-offs and to leverage on synergies between sectors to enhance water, energy and food security.

The changing patterns of energy demand combined with the desire for secure, accessible and environmentally sustainable supply options means that the energy sector is undergoing a transformation led by the rapid adoption of renewables.

References

Abahussain, A.A., Abdu, A.S., Al-Zubari, W.K., El-Deen, N.A. and Abdul-Raheem, M. 2002. Desertification in the Arab region: analysis of current status and trends. *Journal of Arid Environments* 51(4): 521–545.

Abdel-Dayem, S. and McDonnell, R. 2012. Water and food security in the Arab region. pp. 285–301. *In*: Choukr-Allah, R., Ragab, R. and Rodriguez-Clemente, R. (eds.). Integrated Water Resources Management in the Mediterranean Region: Dialogue Towards New Strategy. Springer, New York, London.

ACSAD. 2008. Potential Land use Map of Sudan—Phase 1: Eastern part. Published in the Arab Center for the Studies of Arid Zones and Dry Lands, Final Report (in Arabic).

ADB. 2013. Thinking about water differently: Managing the water–food–energy nexus. (Asian Development Bank (ADB), Mandaluyong City, Philippines).

AFED. 2009. Arab Environment: Climate Change—Impact of Climate Change on Arab Countries, Arab Forum for Environment and Development (AFED).

AFED. 2010. Water: Sustainable Management of a Scarce Resource. El-Ashry, M., Saab, N. and Zeitoon, B. (eds.). Arab Forum for Environment and Development (AFED), Lebanon.

AFED. 2011. Green economy: Sustainable transition in a changing Arab World. *In*: Abaza, H., Saab, N. and Zeitoon, B. (eds.). Arab Environment 4. Arab Forum for Environment and Development (AFED), Beirut, Lebanon.

AFED. 2013. Sustainable energy: prospects, challenges, opportunities. *In*: Gelil, I.A., El-Ashry, M. and Saab, N. (eds.). Arab Environment Annual Report. 6. Arab Forum for Environment and Development, Beirut, Lebanon.

Ahmad, M.H., Azhar, U., Wasti, S.A. and Inam, Z. 2005. Interaction between population and environmental degradation. *The Pakistan Development Review* 44(4): 1135–1150.

Al-Awar, F., Abdulrazzak, M.J. and Al-Weshah, R. 2010. Water ethics perspectives in the Arab region. pp. 29–37. *In*: Brown, P.G. and Schmidt, J.J. (eds.). Water Ethics: Foundational Readings for Students and Professionals. Island Press, Washington, Covelo, London.

AMF. 2010. Statistical Bulletin of Arab Countries, Arab Monetary Fund (AMF), Abu Dhabi.

AWC. 2004. 3nd Arab Water Forum: Final Forum Report. (http://arabwatercouncil.org/3rdAWF/files/report.pdf).

AWC. 2011. 2nd Arab Water Forum: Final Forum Report.

Bachellerie, I.J. 2012. Renewable Energy in the GCC Countries: Resources, Potential, and Prospects, Gulf Research Center.

Barbier, E.B. 2004. Water and economic growth. *The Economic Record* 80(248): 1–16.

Bazilian, M., Rogner, H., Howells, M., Hermann, S., Arent, D., Gielen, D., Steduto, P., Mueller, A., Komor, P., Tol, R. and Yumkella, K. 2011. Considering the energy, water and food nexus: Towards an integrated modelling approach. *Energy Policy* 39: 7896–7906.

Beck, M.B. and Walker, R.V. 2013. On water security, sustainability, and the water-food-energy-climate nexus. *Front. Environ. Sci. Eng.* 7(5): 626–639.

Beddington, J. 2009. Food, energy, water and the climate: A perfect storm of global event? *In*: Lecture to Sustainable Development London.

Bizikova, L., Roy, D., Swanson, D., Venema, H.D. and McCandless, M. 2013. The water–energy–food security nexus: Towards a practical planning and decision-support framework for landscape investment and risk management. The International Institute for Sustainable Development (IISD), Winnipeg, Manitoba, Canada.

Breisinger, C., Ecker, O., Al-Riffai, P. and Yu, B. 2012. Beyond the Arab Awakening: Policies and Investments for Poverty Reduction and Food Security, International Food Policy Research Institute (IFPRI), Washington, DC.

Bruch, C., Jensen, D., Nakayama, M., Unruh, J., Gruby, R. and Ross Wolfarth. 2008. Post-conflict peace building and natural resources. *Yearbook of International Environmental Law* 19: 70–73 and 80–82.

Daher, B.T. 2012. Water, energy, and food nexus: A Basis for strategic planning for natural resources. pp. 221. *In*: Engineering. Purdue University, West Lafayette, Indiana.

Droogers, P., Immerzeel, W.W., Terink, W., Hoogeveen, J., Bierkens, M.F.P., van Beek, L.P.H. and Debele, B. 2012. Water resources trends in Middle East and North Africa towards 2050. Hydrol. Earth Syst. Sci. 16: 1–14. www.hydrol-earth-syst-sci.net/16/1/2012/doi:10.5194/hess-16-1-2012.

El-Katiri, L. and Fattouh, B. 2011. Energy Poverty in the Arab World: The Case of Yemen, Oxford Institute for Energy Studies.

El-Naser, H. 2013. Arab Spring and Water Governance. *In*: Arab Water Week 2013, Arab Countries Water Utilities Association (ACWUA), Amman, Jordan.

Elasha, B. 2010a. Adaptation to Climate Change: Options and Good Practices for the Arab Region, UNDP.

Elasha, B. 2010b. Mapping of Climate Change Threats and Human Development Impacts in the Arab Region, Arab Human Development Report 2010: Research Paper Series-UNDP Regional Burear for the Arab States.

Emmanuel, R. and Baker, K. 2012. Carbon Management in the Built Environment, Routledge.

Erian, W.F., Fares, F.S., Udelhoven, T. and Katlan, B. 2006. Coupling Long-term NDVI for Monitoring Drought in Syrian Rangelands. The Arab Journal for Arid Environments 1: 77–87, Published by ACSAD.

Erian, W. 2010. Drought Vulnerability in the Arab Region: Case Study Drought in Syria-10 Years of Scarce Water (2000–2010), The League of Arab States–Arab Center for the Studies of Arid Zones and Dry Lands (ACSAD).

European Report on Development. 2012. Confronting Scarcity: Managing Water, Energy and Land for Inclusive and Sustainable Growth (European Commission, Overseas Development Institute (ODI), European Centre for Development Policy Management (ECDPM), German Development Institute (GDI/DIE), Brussels, Belgium).

FAO, AQUASTAT. 2015.

FAO. 1996. Rome Declaration on World Food Security and World Food Summit Plan of Action. Rome: Food and Agriculture Organization of the United Nations.

FAO. 2003. Trade Reforms and Food Security: Conceptualizing the Linkages. Rome: Food and Agriculture Organization of the United Nations. ftp://ftp.fao.org/docrep/fao/005/y4671e/y4671e00.pdf.

FAO. 2006. World agriculture towards 2030/2050, Rome, Italy: FAO, Food and Agriculture Organization of the United Nations, Rome, Italy.

FAO. 2007. Coping with water scarcity challenge of the twenty-first century (http://www.fao.org/3/a-aq444e.pdf.

FAO. 2008. Twenty-Ninth FAO Regional Conference for the Near East. Egypt, Food and Agriculture Organization of the United Nations.

FAO. 2009. Irrigation in the Middle East Region in figures: AQUASTAT survey—2008. Rome: Food and Agriculture Organization of the United Nations.

FAO. 2009. World Summit on Food Security–Declaration of the World Summit on Food Security. Rome: Food and Agriculture Organization of the United Nations.

FAO. 2011a. The state of the world's land and water resources for food and agriculture (SOLAW)–Managing systems at risk. Food and Agriculture Organization of the United Nations, Rome and Earthscan, London.

FAO. 2011b. Energy-smart food for people and climate. Issue Paper. Rome: Food and Agriculture Organization of the United Nations.

FAO. 2012a. Coping with water scarcity An action framework for agriculture and food security. Rome: Food and Agriculture Organization of the United Nations.

FAO. 2012d. Mutually acceptable mechanism on integrated use of water resources in Central Asia. Application of the scenario approach. Ankara: Food and Agriculture Organization of the United Nations.

FAO, IFAD and WFP. 2014. The State of Food Insecurity in the World 2014. Strengthening the enabling environment for food security and nutrition. Rome, FAO.

FAO. 2014. The Water-Energy-Food Nexus: A new approach in support of food security and sustainable agriculture. Rome. http://www.fao.org/3/a-bl496e.pdf.

FAO. 2011. Energy-Smart Food for People and Climate. FAO, Rome. http://www.fao.org/docrep/014/i2454e/i2454e00.pdf.

Food-Energy-Ecosystems Nexus for Reconciling Different Uses in Transboundary River Basins– UNECE Water Convention Draft Methodology. (UNECE, Geneva, Switzerland).

Forbes. 2008. Virtual Water. http://www.forbes.com/2008/06/19/water-food-trade-tech-water08-cx_fp_0619virtual.html. Accessed on February 12, 2014.

Gober, P. 2010. Desert urbanization and the challenges of water sustainability. *Current Opinion in Environmental Sustainability* 2: 144–150.

Gonzalez, C.G. 2004. Trade liberalization, food security, and the environment: the neoliberal threat to sustainable rural development. *Transnational Law & Contemporary Problems* 14(2): 419–498.

Gregory, P.J., Ingram, J.S.I. and Brklacich, M. 2005. Climate change and food security. *Phil. Trans. R. Soc. B* 360: 2139–2148.

Habib-Mintz, N. 2009. Mapping Climate Change Issues, Initiatives, and Actors in the Arab Region, This report was produced as a part of policy research initiative undertaken by the Regional Bureau of Arab Region of UNDP for the UNFCCC's Copenhagen meeting on climate change in 2009.

Haddadin, M.J. 2003. Exogenous water: A conduit to 101 globalization of water resources. *In*: Hoekstra, A.Y. (ed.). Virtual Water Trade: 102 Proceedings of the International Expert Meeting on Virtual 103 Water Trade. UNESCO-IHE, Delft, The Netherlands.

Hoff, H. 2011. Understanding the Nexus. Background Paper for the Bonn2011 Conference: The Water, Energy and Food Security Nexus. tockholm Environment Institute, Stockholm. *In*: Bonn2011 Conference, Stockholm Environment Institute (SEI), Germany.

Hoff, H., Fielding, M. and Davis, M. 2013. Turning vicious cycles into virtuous ones. *Rural* 21(03): 10–12.

Homer-Dixon, T.F. 1994. Environmental scarcities and violent conflict: evidence from cases. *International Security* 19(1): 5–40.

Howe, C.W. 1976. Effects of water resource development on economic growth: the conditions for success. *The Natural Resources Journal* 16: 939–955.

Idriss, A., Lupien, J., Sheet, I. and Romanos, H. 2012. Food Security, the Moving Borders of Poverty, Free Markets and Political Interventions, Middle East North Africa Food Safety Associates, pp. 18.

ICIMOD. 2012. Contribution of Himalayan ecosystems to water, energy, and food security in South Asia: A nexus approach. (International Centre for Integrated Mountain Development (ICIMOD), Kathmandu, Nepal).

IPCC. 2007. Summary for policymakers. *In*: Solomon, S., Qin, D., Manning, M., Chen, Z., Marquis, M., Averyt, K.B., Tignor, M. and Miller, H.L. (eds.). Climate Change 2007: The Physical Science Basis. Contribution of Working Group I to the Fourth Assessment Report of the Intergovernmental Panel on Climate Change. Cambridge University Press, Cambridge, UK and New York.

IPCC. 2013. Summary for policymakers. pp. 27. *In*: Stocker, T.F., Qin, D., Plattner, G.-K., Tignor, M.M.B., Allen, S.K., Boschung, J., Nauels, A., Xia, Y., Bex, V. and Midgley, P.M. (eds.). Climate Change 2013: The Physical Science Basis. Contribution of Working Group I to the Fifth Assessment Report of the Intergovernmental Panel on Climate Change, Intergovernmental Panel on Climate Change (IPCC). Cambridge, United Kingdom and New York, NY, USA.

IRENA. 2015. Renewable Energy in the Water, Energy & Food Nexus.

Kfouri, H. 2013. Food security and policy streams theory in the middle east. *The Wagner Review*.

Khatib, H. 2010. The Water and Energy Nexus in the Arab Region. League of Arab States, Cairo.

Khoday, K. 2011. Sustainable development & green economy in the Arab Region. *In*: Arab Development Challenges Background Paper 2011, UNDP, Cairo, Egypt.

Larson, D.F. 2013. Blue Water and the Consequences of Alternative Food Security Policies in the Middle East and North Africa for Water Security, The World Bank.

Meisen and Tatum. 2011. The Water-Energy Nexus in the Jordan River Basin: The Potential for Building Peace through Sustainability. By GENI: Global Energy Network Institute. (https://www.geni.org/globalenergy/research/water-energy-nexus-in-the-jordan-river-basin/the-jordan-river-basin-final-report.pdf).

Ministry of Water and Irrigation (MWI). 2011. Annual Water Budget Report (2011), Amman, Jordan.

Mirkin, B. 2010. Arab Human Development Report 2010–Research Paper Series–Population Levels, Trends and Policies in the Arab Region: Challenges and Opportunities, UNDP.

MOE. 2012. The Energy Strategy for Ministry of Energy 2012, Amman, Jordan.

Mohtar, R. and Daher, B. 2012. Water, energy, and food: the ultimate nexus. *In*: *Encyclopedia of Agricultural, Food, and Biological Engineering*.

Molden, D. (ed.). 2007. Water for life, water for good: A comprehensive assessment of water management in agriculture. International Water Management Institute, Columbo and Earthscan, London.

(MWI) Ministry of Water and Irrigation. 2011. Annual Water Budget Report (2011), Amman, Jordan.

MWI. 2013. Jordan Water Sector Facts and figures 2013. Ministry of water and Irrigation 2013.

OECD. 2013. Renewable Energies in the Middle East and North Africa: Policies to Support Private Investment, OECD Publishing.

Philips, D., Daoudy, M., McCaffrey, S., Ojendal, J. and Turton, A. 2006. Transboundary Water Cooperation as a Tool for Conflict Prevention and for Broader Benefit-Sharing, Ministry of Foreign Affairs, Sweden, Stockholm, pp. 15.

QEERI. 2012. Nexus 1.0. A Resource Management Strategy Guiding Tool.

Richards, A. 2001. Coping with water scarcity: the governance challenge. *Center for Global, International & Regional Studies* 1–67.

Shetty, S. 2006. Water, Food Security and Agricultural Policy in the Middle East and North Africa Region, The World Bank.

Sowers, J., Vengosh, A. and Weinthal, E. 2011. Climate change, water resources, and the politics of adaptation in the Middle East and North Africa. *Climatic Change* 104: 599–627.

Starr, J.R. 1993. Water wars. *In*: Bulloch, J. and Darwish, A. (eds.). Water Wars: Coming Conflicts in the Middle East, London.

Tignino, M. 2010. Water, international peace, and security. *International Review of the Red Cross* 92(879): 647–674.

UN. 2009. Oil in a Low-carbon Economy, UN Chronicale–The Magazine of the United Nations.

UN-HABITAT. 2012. The State of the Arab Cities 2012: Challenges of Urban Transition, United Nations Human Settlements Programme (UN-Habitat), pp. 232.

UNDESA. 2011. World Population Prospects, the 2010 Revision, United Nations Department of Economic and Social Affairs, New York. [http://esa.un.org/unpd/wpp/Excel-Data/population].

UNDP. 1994. New dimensions of human security–Human Development Report, United Nations Development Programme, New York.

UNDP. 2003. Water governance and poverty reduction: draft practice note for the bureau for development policy, energy and environment group, United Nations Development Programme, New York.

UNDP. 2004. Water Governance for Poverty Reduction. Key Issues and the UNDP Response to Millennium Development Goals. http://www.ungei.org/infobycountry/files/UNDP_Water_Governance.pdf, United Nations Development Programme, New York.

UNDP. 2006. Human Development Report, Beyond Scarcity: Power, Poverty and the Global Water Crisis, 203.

UNDP. 2009. Development Challenges for the Arab Region: Food Security and Agriculture.

UNDP. 2011. Arab Development Challenges Report 2011: Towards The Developmental State in the Arab Region, United Nations, New York, USA.

UNDP. 2013a. The Rise of the South: Human Progress in a Diverse World. *In*: Human Development Report 2013, United Nations Development Programme (UNDP).

UNDP. 2013b. Water Governance in the Arab Region: Managing Scarcity and Securing the Future, 182.

UNEP. 2013. Arab Region: Atlas of Our Changing Environment, United Nations Environment Programme (UNEP), Nairobi, Kenya.

UN-ESCAP. 2013. Water-Food-Energy Nexus in Asia and the Pacific Region. Discussion Paper. (UN-ESCAP, Bangkok, Thailand).

UNESCO. 2003. Water Security and Peace: A Synthesis of Studies Prepared under the PCCP–Water for Peace Process. Cosgrove, W.J. (ed.). UNESCO-Green Cross International Initiative, pp. 9–18.

UNESCO. 2012. The United Nations World Water Development Report 4: Managing Water under Uncertainty and Risk: Executive Summary, United Nations Educational, Scientific and Cultural Organization, Paris, France.

UNESCO. 2012. World Water Assessment Programme (WWAP), World Water Development Report, Vol. 1: Managing Water under Uncertainty and Risk (Paris: UNESCO, 2012).

UNFCCC. 2012. Renewable Energy Programme of Activities in Middle East and North Africa. UNFCCC.

UNFCCC. 2014. Jordan's Third National Communication on Climate Change. Jordan: UNFCCC.

UN-ESCWA and BGR (United Nations Economic and Social Commission for Western Asia; Bundesanstalt für Geowissenschaften und Rohstoffe). 2013. Inventory of Shared Water Resources in Western Asia. Beirut. https://waterinventory.org/sites/waterinventory.org/files/chapters/00-shared-water-resouces-in-western-asia-web.pdf.

Verner, D. 2012. Adaptation to a Changing Climate in the Arab Countries: A Case for Adaptation Governance and Leadership in Building Climate Resilience. Verner, D. (ed.). World Bank, Washington, DC: World Bank, pp. 402.

Verner, D. and Breisinger, C. 2013. Economics of Climate Change in the Arab World: Case Studies from the Syrian Arab Republic, Tunisia, and the Republic of Yemen, World bank, Washington DC.

Verner, D., Lee, D.R., Ashwill, M. and Wilby, R. 2013. Increasing Resilience to Climate Change in the Agricultural Sector of the Middle East: The Cases of Jordan and Lebanon, The World bank, Washington, D.C.

Vörösmarty, C.J., Green, P., Salisbury, J. and Lammers, R.B. 2000. Global water resources: vulnerability from climate change and population growth. *Science* 289: 284–288.

Waughray, D. 2011. Water Security: The Water-Energy-Food-Climate Nexus. The World Economic Forum water initiative/Edited by Dominic Waughray.

WB. 2007. Middle East and North Africa (MENA) Sustainable Development Sector Department (MNSSD) Regional Business Strategy to Address Climate Change, The World Bank, Washington DC.

WEF. 2011. Global Risks 2011. 6th Edition. (World Economic Forum (WEF), Cologne/ Geneva).

WEF. 2011a. Water Security: The Water–Food–Energy–Climate Nexus. Island Press, Washington.

WorldBank. 2013. The World Bank Data Base , access June 2015.

WWF6. 2012. Theme 2.2. Contribute to Food Security by Optimal Use of Water", Core Group Session Proposal. *In*: 6th World Water Forum, Marseille, France.

www.fao.org/nr/water/, viewed June26, 2015 data for 2014.

Zeitoun, M. and Mirumachi, N. 2008. Transboundary water interaction I: reconsidering conflict and cooperation. *International Environmental Agreements: Politics, Law and Economics* 8(4): 297–316.

Chapter-7

Recognizing the Food-Energy-Water Nexus

A Renewed Call for Systems Thinking

Victoria Pebbles

INTRODUCTION

Food, energy, water: it's the golden triangle that supports all life. At its most basic level, energy is foundational to everything. But in this chapter, energy refers to the energy systems we create as a human society to give us power, heat, and transportation that moves us from place to place. Food systems refer to the way we grow, cultivate, process, and deliver the food we eat. Water supports food and energy, energy is needed to create food and deliver clean water, and food is essential for basic human survival.

Ignore any one of these and the other two falter. That seems obvious, but a closer look at how human society, and North America in particular, reveals how we've been chipping away at this foundation at our own peril. This chapter offers a case study of how this has played out in the Great Lakes region.

The Great Lakes *region* is generally defined as the geography of the eight U.S. states and two Canadian provinces that border one or more of the Great Lakes.[1] By way of reminder, those lakes are: Superior, Michigan, Huron, Erie and Ontario, or their connecting channels. The Great Lakes *basin* is a subset of that region—the watershed—that drains into one of the Great Lakes or their connecting channels.

Program Director, Great Lakes Commission, 2805 S. Industrial Hwy. Suite 100, Ann Arbor, MI 48104.

[1] U.S. States: Illinois, Indiana, Michigan, Minnesota, New York, Ohio, Pennsylvania, Wisconsin. Canadian provinces: Ontario and Quebec.

Most of the references in this chapter refer to the larger region which is the political economy that drives the actions that occur in and influence energy, water, and agricultural systems in the basin. The term *basin* is used where it is important to convey the watershed as an ecosystem of highly interdependent living organisms in conjunction with the nonliving components of their environment (e.g., air, water and mineral soil), interacting as a system and linked together through nutrient cycles and energy flows.[2]

About 97 percent of all the water on earth is salt water. Of the remaining three percent that is freshwater, the Great Lakes constitute about 20 percent. In other words, one fifth of the earth's fresh surface water resides in the Great Lakes of Superior, Michigan, Huon, Erie and Ontario. That's a lot of fresh water. To put it in perspective, if you dumped out the lakes, they would cover the continental U.S. in nearly 10 feet of water.

Thirty-six million people depend on Great Lakes for drinking water in the US and Canadian sides of the Great Lakes Basin, including the Chicago metro area (GLC, 2013). Forty-four billion gallons of water are used each day to support municipal water (water for homes and businesses), agricultural (i.e., food) production, and industrial use, including the power sector (GLC, 2013). The Great Lakes Commission—a binational agency that advises the eight states and two provinces that border the Great Lakes and the St. Lawrence River tracks water use by sector in the binational Great Lakes region using categories of users consistent with those used by the U.S. Geological Survey—a part of the U.S. Department of the Interior—which tracks water use across the United States.

Water for Agriculture

(Agriculture's Dependence on Water; Impacts on Water Resources; and Implications of Past and Current Policy and Management Approaches)

A brief history of agriculture in the great lakes region

Agriculture, the industry that produces food and fiber, has transformed almost exponentially over the past century and a half. We still like to think of families that weathered innumerable hardships to clear and cultivate land and grow the food and fiber that allowed communities to establish, survive and ultimately flourish. The Great Lakes region was largely covered by hardwood forests in the north, but the south had fewer trees, was flat, and was endowed with nutrient rich black soil created by vast swamps. Today these are also known as wetlands. The flat land, rich soils offered great appeal for growing crops, but there was often too much water that presented a risk to crops. As a result, the land had to be drained.

In the 1800s and 1900s as the Great Lakes region's industrial development also included development of the agricultural industry. Rapid, large-scale clearing of land occurred; agriculture stripped soils of vegetation and soil washed away to the lakes;

---

[2]   Excerpted from Wikipedia. February 22, 2017.

sediment coming off these lands clogged tributaries and silty deltas and altered the flow of the rivers. Habitats supporting fish and other creatures that depended on these wetlands were severely degraded or destroyed. Little to no deep-rooted vegetation (i.e., trees) to soak up the water meant more rain washed off croplands, increasing seasonal flooding, streambanks erosion, and further deterioration of aquatic habitats, including those at the mouths of rivers and ultimately along the Great Lakes nearshore.

The Great Black Swamp, a vast wetland area that once occupied what is now much of northwest Ohio and parts of southwest Michigan, covered an estimated 1,500 square miles (4,000 km²) from what is today Sandusky, Ohio to Monroe, Michigan was completely cleared and drained.[3,4] An equally vast system of trenches were dug to ensure that excess water was drained off the fields with little thought to where that water would go. So long as water wasn't sitting on the fields, drowning the crops, it was a good thing. Given the region's long hard winters, many farmers complemented their crops with livestock that could make up relatively short growing season, ensuring an agricultural livelihood.

Over time, competition drove farmers to intensify operations, and advances in technology and transportation provided access to larger markets, further encouraging shifts towards mechanical operations. Some of the technological advances were in farming equipment. However, advances in the chemical industry allowed farmers to grow more of the same few crops that would guarantee a more stable income. Corn, soybeans, and wheat were always in demand and made a superior match for the soil conditions and growing season offered in the southern Great Lakes region.

Chemicals made to keep unwanted plants at bay (herbicides) and to keep unwanted insects at bay (pesticides) were part of the technology revolution that greatly enhanced farmers' ability to focus more intensively on just a few crops. The unintended consequence was that the water and soil coming off the fields became laden with residual chemicals from the herbicides and pesticides. Advances in the chemical industry also brought the advent of chemical fertilizers. While traditional farming had used manure and crop residues to help nourish the soil, a more precise application of specific nutrients, could improve yields. Whether the crops needed all those nutrients, was not questioned until the 1960s when Lake Erie became fouled with massive algae blooms.

Scientists discovered that excessive phosphorus in the runoff was a primary culprit and policy-makers responded over the next couple of decades with regulations and incentives to reduce phosphorus in household soap products and, to a lesser extent, residential fertilizers. But industrial agriculture was left largely untouched.

Local consequences of global agribusiness

As of the 2010s, approximately 36 percent of the Great Lakes basin land area is used for agriculture.[5] Agriculture is a leading industry in every GL state and

---

[3] Dempsey, D. 2001. Ruin and Recovery: Michigan's Rise as a Conservation Leader. University of Michigan Press. Ann Arbor.
[4] The Great Black Swamp, Wikipedia. Downloaded March 27, 2017.
[5] Reference needed.

Ontario leads agricultural production in Canada. The region's agricultural profile is primarily livestock, including dairy, and commodity crops (corn, wheat, soybeans) with some niche and seasonal markets for specialty fruits and vegetables. Advances have been made to protect water resources from industrial polluters, but similar approaches to address pollution from agricultural sources have been resisted. The societal values that came to honor hardworking farmers that toiled to grow food and fiber has remained stuck in time, while farmers and farming practices have utterly transformed from the iconic family farm that we cling to. In reality, most of those family farms no longer exist. The famous *American Gothic* painting by Grant Wood[6] featuring an iconic scene of the ma and pa farmer with a pitchfork is emblazoned into popular psyche, but in 2017, those family farms are few and far between. Family farms have been virtually replaced by large business farms that are owned or heavily influenced by chemical companies, seed companies, and conglomerates (Melanie et al., 2011).[6a] People working in and around agriculture are no longer even referred to as farmers, but as "agricultural producers". Language has also mechanized as farmers have become "agricultural producers" whose products ultimately get use or eaten by "consumers".

The rise of organic farming, farmers markets, niche markets for specialty crops and products (e.g., Michigan cherry jam) coupled with increased consumer awareness about the quality of the food they eat and the conditions that produced that food are slowly helping to create new, more sustainable markets regionally and locally. Globalization—the well-established markets and demand for corn, soybeans and wheat in Asia and Africa and South America—continues to drive the production of major commodity crops across the Great Lakes region.

However, the organic movement only goes part way down the supply chain—to the pesticides and herbicides and, to some extent, chemical fertilizers. It's a great start, but more is needed to also consider the interdependence of agriculture, water, and energy.

Most of the food grown in America's "breadbasket, or which the Great Lakes region is a part, is not the food consumed by local consumers. The vast majority is one of three crops: corn, wheat, and soybeans. Field after field is sown with single or "mono culture crops". Most of the corn is not the sweet kind people of the Great Lakes region savor with butter and salt at an August barbeque. It goes into feed for cows, and is made into various products; the most commonly known is probably high fructose corn syrup. Some of the wheat grown in the region is used for flour for breads and pastries that people in the Great Lakes region eat regularly, but the vast majority of wheat grown in the Great Lakes region doesn't end up in local breads or Aunt Mary's pumpkin muffins. Like corn and soybeans, most wheat grown in the Great Lakes region is shipped to other parts of the globe as part of the global commodities market.[7] Apart from tomatoes grown in southeast Ontario and

---

[6]   *American Gothic* is a painting by Grant Wood in the collection of the Art Institute of Chicago. Painted in 1930, it depicts a farmer standing beside a woman that has been interpreted to be either his wife or his daughter. Source; Wikipedia download February 22, 2017.

[6a] Melanie J. Wender, Goodbye Family Farms and Hello Agribusiness: The Story of How Agricultural Policy is Destroying the Family Farm and the Environment, 22 Vill. Envtl. L.J. 141 (2011). Available at: http://digitalcommons.law.villanova.edu/elj/vol22/iss1/6.

[7]   U.S. Census Bureau, Statistical Abstract of the United States, 2012. Chapter 17: Agriculture.

seasonal niche fruit markets, the vast majority of the lettuces, broccoli, celery, and other fruits and vegetables that are part of a healthy daily diet are imported from other parts of North America and the world. And they are grown on land owned by large corporations and that rely on genetically modified (pest resistant) seeds owned by multinational corporations. The opportunity costs and externalities associated with modern agribusiness is not well known or understood by the public. It's all too overwhelming to think that the vast majority of food and fiber products available to consumers to support human life are fundamentally unsustainable for life in the broader context.

Too much of a good thing is no good

Plants and animals that we call "agricultural products" do indeed require energy and water to grow and flourish. Traditionally, this took the form of energy from the sun with the aid of microorganisms and nutrients in the soil and water from the rain. Plants were either consumed, or used to feed livestock which were consumed. Over time, societal innovations created ways to process both plants and meat to extend the useful life of agricultural products. Those in the agriculture industry also became more knowledgeable about which specific chemicals and nutrients, or combinations thereof, were needed to get the most plant or animal for the least input—in terms of costs. Scientists have cracked the code and broken down specific needs by plant or animal species, enabling agricultural practices that focus narrowly on producing the desired outcome: namely yields and size. There are huge opportunity costs that society is now bearing at multiple environmental, social and even psychological levels. Take milk for example. Big-udder cows that feed on corn and are hooked up to milking machines produce more milk faster and cheaper than ever before, but their udders are disproportionality large and prone to illness; animal suffering is little acknowledged at best. These cows never see the light of day and are valued only for their milk—not their hide or their meat or the various other parts of their organism, which are largely treated as by products and used for cheap dog food or are simply discarded. More poignant to the Great Lakes and their tributary rivers and streams: the volume of manure created by modern livestock operations is so concentrated with nutrients that it presents serious risks to the land and nearby waterways.

In the late 1960s, Lake Erie was so polluted and so badly plagued with algae blooms that it was declared dead. The Lorax, a famous children's story first published in 1971 by the late Dr. Seuss chronicles the plight of the environment and the Lorax, who speaks for the trees against the greedy Once-ler, who ravages the environment in pursuit Truffula Trees. In it, he writes:

> *"you're glumping the pond where the Humming-Fish hummed! No more can they hum for their gills are all gummed. So, I'm sending them off,. Oh, their future is dreary. They'll walk on their fins and get woefully weary in search of some water that isn't so smeary. I hear things are just as bad up in Lake Erie."*

The line "I hear things are just as bad up in Lake Erie" was removed more than 14 years after the story was published after two research associates from the

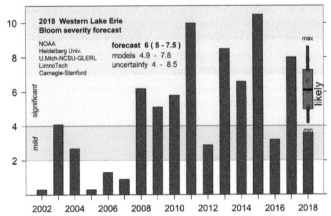

Figure 7.1 Bloom severity index for 2002–2018, and the forecast for 2018. The index is based on the amount of biomass over the peak 30-days. The 2018 bloom had a severity of 3.6, comparable to 2016 (3.2). 2011 had a severity of 10; 2015 was 10.5. Source: Extracted from NOAA, 2018.

Ohio Sea Grant Program wrote to Seuss about the clean-up of Lake Erie.[8] Following extensive phosphorus reduction efforts initiated in the 1970s, harmful algal blooms in Lake Erie were largely absent. However, blue-green algae (cyanobacteria) blooms began to reappear in the western basin of Lake Erie in the mid-1990s. Forty years after initial efforts to reduce phosphorus entering the Great Lakes, toxic algae blooms came back with a vengeance. The 2011 harmful algae bloom in western Lake Erie was the largest bloom in the history of the Great Lakes until it was surpassed just three years later by an even larger more severe bloom in 2015 (Figure 7.1).[9]

The moist, nutrient-rich soil that fueled the success of agriculture in the southern Great Lakes region has become a type of saturated dirt. It has plenty of nutrients, but is often deficient in essential microorganism that can constitute a healthy soil and that can help use, distribute or convert excess nutrients to deliver the full range of what pleats need. It's as if we realized vitamin C is good to for a healthy immune system so we only focused on giving people vitamin C. Since it is water soluble, vitamin C in excess is generally not toxic to humans but in excess can cause short term and even long term health problems.[10] Well, we know that vitamins and minerals behave differently depending on their environment—the system—in which they are introduced. It is no different with phosphorus. Too much of it isn't toxic to the plants, but it contributes to excessive plant growth (including algae) that causes imbalances in water quality and creates conditions for toxic algae blooms. Too much of a good thing is no good.

Water and climate dependency

Agriculture is a leading industry and a driving economic force in the southern part of the Great Lakes region. The agriculture sector has traditionally relied on rain water to

---

[8]   Morgan and Morgan. 1995. Dr. Seuss & Mr. Geisel: A Biography. Random House. Da Capo Press, Incorporated, 2009. ISBN 978-0-679-41686-9.

[9]   Experimental Lake Erie Harmful Algal Bloom Bulletin. 26 October 2018, Seasonal Assessment.

[10]   Consumerlabs.com, accessed February 22, 2017.

feed crops, but that pattern is likely to shift with climate change as weather patterns rains become less predictable. The region is already seeing shifts to more irrigation, which relies on surface and or ground water. And climate experts also predict more frequent and more severe storms, which means when it does rain, there are increasing risks of having too much water, thereby augmenting the need for systems like drain tiles to ensure water drains off fields before drowning the crops. The paradox is that increased efforts to better manage water coming out of drain tiles is not likely to be able to keep pace with water management needs that are arising and will continue with climate change.

Another factor worth considering is that when crops relied on rainfall, that water was not removed from rivers, streams or groundwater, so there was no concern about water use for agriculture having much of an impact on water supply. Again, under climate change, the situation it totally different. Irrigated systems can be used to deliver water more reliably in time and in amount, but that water must come from rivers, streams or underground. The laws, policies and institutions in the region are not well-equipped to manage large new uses of surface or groundwater. The legal framework allows landowners to draw water from adjacent rivers or underground sources as long as those withdrawals are reasonable and do not interfere with the ability of others to also use those water sources. This approach, known as riparian rights doctrine in the U.S., predominates east of the Mississippi in the U.S., including the entire U.S. side of the Great Lakes Basin. The riparian approach seems reasonable where water supplies are consistent and plentiful and it has worked relatively well as long as rains could be relied on to nourish crops. However, this approach is not sustainable over the long term, where rains are no longer reliable. In some parts of the region farmers are looking to ground and surface water supplies for consistent reliable water sources, creating additional pressure on water sources that are already at risk due to insufficient groundwater infiltration, or seasonally low flows, or both (GLC Water Energy Nexus, 2011). This is likely to become more common in the future. New policies and governance models are needed to help the region respond to the trifecta of increasing demand on water for agricultural crops, the attendant "loss" of that water since most of water used for agriculture is consumed by the plants and not returned to surface or groundwater sources, and the accelerated rate and increased quantity of water coming off farms (either over the surface or through drain tiles) during storm events.

Putting the pieces back together

The Great Lakes region is part of North America's powerhouse of scientific, social and political institutions that were built on breaking things into classes and compartments, analyzing them and managing them or putting them back together. From that we have highly esteemed scientific methods, scientists that interface with a highly complex network of policies, programs, and social and cultural norms also based on grouping, classifying, dissecting and compartmentalizing. This has offered tremendous advances to North American society. Along the way, however, there has been less and less emphasis on the system that represents the collective whole. Environmental impacts are too often viewed, experienced, and responded to as a series of seemingly isolated events. But they are not. They are connected and will ultimately have a cascading effect leading to multiple collapses of seemingly varied, yet deeply interconnected systems, like water and agriculture and energy.

This epic failure to recognize how water, food and energy are connected systems and are, indeed, connected nested critical elements supporting broader societal needs is unsustainable and, if not corrected, could ultimately cause total societal collapse.[11] In his book, Collapse, Jared Diamond's five-point framework illustrates common elements shared by societies that have collapsed throughout human history. Two of them are environmental damage and climate change. The other three—loss of friendly contacts, rise of hostile contacts, and dysfunctional political and cultural practices—are distinct but not unrelated. Indeed, Diamond admits that he is unaware of cases where environmental damage alone has caused a society's collapse. However, the inability to perceive environmental risks and properly address them through corrective measures including appropriate political and cultural changes has.[12]

Energy

(Energy's Dependence on Water: Impacts on Water and Implications Past and Current Policy and Management Approaches)

Merriam-Webster defines energy as a fundamental entity of nature that is transferred between parts of a system in the production of physical change within the system and usually regarded as the capacity for doing work. With this definition, it is easy to see how energy is a system that is foundational to life on earth and indeed how the cosmos works. Merriam-Webster also defines energy as "usable power (as heat or electricity) and the resources for producing such power." This chapter focuses on the latter form of energy—as power—and its interrelationship to water and food. Interestingly, the overarching point of this chapter is aligned to the broader definition that recognizes the inherent energy in all things and their connection to one another.

The generation of power or fuel required to heat and cool homes and businesses, or to move cars, trucks, trains and ships is essential for modern economies to function. An extraordinary interdependence between energy and water underlines the modern Great Lake region's economy. Much like water when it's in pipes, energy is part of a vast grid that most people recognize only when they turn a knob or flip a switch, and only appreciate when after that knob or switch has stopped working. The intricacies of how that water comes out of the faucet, or how that light turns on, are generally lost on the public. It's not a criticism of the public, but more a function of human nature captured by an age-old adage: out of sight, out of mind.

We rely on electric energy every day for almost all of our personal and professional or business activities. The two forms of electric power produced that supply the vast majority of electricity in the Great Lakes region are water-based. First is hydro-electric power where the natural gravitational forces of the earth enable

---

[11]   Diamond, J. 2005. Collapse: How Societies Choose to Fail or Succeed. Penguin Books. New York.
[12]   Ibid. In particular, see Chapters 2, 4, and 5 on Easter Island, the Anasazi, and the Maya, respectively.

capturing that falling water to turn a turbine and create electric power. Hydropower is predominant in the northern parts of the Great Lakes region, including parts of Minnesota, New York, Quebec, and Ontario where the natural topography allows it. [Hydroelectric power requires tremendous amounts of water, and while the land and local habitats around that water are undoubtedly subject to disturbances, the water itself is not altered in quantity or quality. To be sure, damning a river system to create hydroelectric power causes ecological disturbances and brings potential for attendant water quality problems upstream and down from those dams.] The second type is thermo-electric power, where some type of fuel (typically coal or natural gas, but also nuclear reactor) creates heat to boil water and create steam that turns a turbine and voila—electric power! The southern, flatter part of the Great Lakes region relies primarily on thermoelectric power. Almost 90 percent of the power generation on the U.S. side of the Great Lakes basin comes from thermoelectric power.[13] Together hydroelectric power and thermo-electric power use represent the more than 95 percent of the water use in the Great Lakes basin.[14] However, hydropower does not take water out of the system and is often not considered a water "use." If we leave out hydropower, thermo-electric power is still the largest water user by sector, representing more than 70 percent of the water used, or withdrawn, from the Great Lakes every day. Because of the predominant technology, known as "once-through cooling" most of the water used for power generation is returned to the system (although usually at higher temperatures). However, every type of energy production has environmental consequences and power generation is no exception. The alternative to once-through cooling systems is a "closed loop" system which takes less water out of lakes and rivers, but consumes much more water meaning that less is actually returned to the source. The power sector's reliance on water where a lot is withdrawn but most is put back sits in contrast to agriculture where, at least historically not much water is taken out of the system, but what is used is lost to the system because it becomes part of the plant or animal (or vapors through evapotranspiration).

Most of the water used for thermo-electric power in the Great Lakes basin comes directly from one of the five Great Lakes where water supplies are vast and generally cold and the risks to water quality or quantity are minute.[15] However, about 25 percent of the thermo-electric power plants in the Great Lakes Basin take water from a river system or groundwater. In these instances, where there is less water in quantity and the relative pollution loads into those waters are generally greater as is the demand to use the water, there are greater risks of water use conflicts—over supply and quality. A 2011 report by the Great Lakes Commission identified looked closely at power generation on the U.S. side of the Great Lakes Basin and potential

---

[13] Great Lakes Commission. 2011. Integrating Energy and Water Resources Decision Making in the Great Lakes Basin: An Examination of Future Power Generation Scenarios and Water Resource Impacts. Ann Arbor, Michigan.

[14] Great Lakes Regional Water Use Data Base; Summary of Water Use Across Great Lakes Basin for 2015. Accessed at: https://waterusedata.glc.org/graph.php?type=summary&year=2015&units=gallons. For more information, see: http://waterusedata.glc.org.

[15] There are real risks to aquatic biota through impingement and entrainment from thermo-electric water intakes.

impacts on water quantity and quality. The analysis looked at 102 sub-watersub-watersheds, known in the U.S. as a "HUC-8" scale: the Raisin River in Michigan, the Lower Fox in Wisconsin, and the Lower Genesee in New York fall into this scale.[16] Each watershed was analyzed for three types of threat: (1) inadequate surface water supply during summer seasons; (2) excessive temperature change; (3) overdrawn groundwater supplies. The study also looked at water quality.

That analysis showed that nearly a quarter of the sub-watersheds are vulnerable to having inadequate water supplies during "low flow seasons"—hot summer months when stream flows are at their lowest and demand for air-conditioning (provided through thermo-electric power) peaks. These conditions are likely to be more frequent in the future as the impacts of climate change become more severe. As summer months get hotter with climate change, and there is greater need for water for a variety of purposes, this risk is likely to play out in greater water use conflicts, inadequate water to support aquatic life, or both.

That same study showed that more than half—57 percent—of the sub-watersheds were at moderate to high risk due to changes in water temperature. That is, those cold and cool rivers inhabited by cold-water species are close to a tipping point where minor temperature changes could substantially impact the natural communities of those areas. Thermo-electric power plants generally release water back into rivers at temperatures warmer than when it was withdrawn and sometimes warmer than the fish and other creatures are adapted to. When temperatures get hot, people can turn on the AC. Some communities are even setting up cooling centers for the poor and elderly in the case of power outages and/or extreme heat days. Fish don't have that option. While fish in large lakes and rivers may be able to move to cooler spots, it doesn't mean those cooler areas are ideal habitat.

A look at water quality is similarly discouraging. When it comes to more basic pollution—excessive nutrients, sediments, heavy metals, or chemicals—more than a third of these same watersheds are already considered moderately to highly impaired according to U.S. EPA and state reports. All told, one-fifth of the Great Lakes Basin's sub-watersheds rank *high* for two or more of these risk factors. Although the study focused on the U.S. side of the Great Lakes, it is reasonable to expect that similar conclusions would be found on the Canadian side of the lakes.

The consideration of water quality in this section on energy is deliberate. Water quantity and water quality are related. If there is a risk of inadequate water supply, that risk is faced both by the humans as well as the fish and wildlife that depend on that water. If people need a new thermal electric power plant in a drought-prone area, who will win: the fish or the people? Are governance systems in place adequate to overcome this false dichotomy? And basic chemistry informs that with more water, pollution is generally diluted. With less water, pollutants are more concentrated. Where there is inadequate water that is also of poor quality, the potential risks and impacts are compounded.

---

[16]   HUC refers to Hydrologic Unit Code established by the U.S. Geological Survey which divides and sub-divides the United States into successively smaller regions based on hydrologic boundaries, or watersheds.

These are early warning signs of current and impending system problems. Even in this vast region of 20 percent of the world's fresh surface water, there are places suffering from poor water quality and at risk of additional problems related to quality and supply if current governance and management approaches are not modified and adapted to recognize the consequences of current governance and management approaches and the compounding impacts of climate change.

Water

(Water as a Socio-Economic Driver, its Reliance on Energy, and Impacts and Implications of Past and Current Policy and Management Approaches)

Water is essential to life. Humans need to drink it to survive as do many other species. Some species need water to breathe. In all contexts, the quality of the water is important. Poor water quality puts the human health, and the health of the other parts of the ecosystem, of which we are part, at risk. Working for a Great Lakes organization for more than 20 years, I've often gotten the question: "So, how are the Great Lakes doing anyway?"

Starting in 2011 when the then-largest toxic algae bloom on record appeared in Lake Erie, there have been several major incidents telling, or screaming, that things are not good. There was the 2014 toxic algae bloom, certainly not the worst ever, but bad enough that it shut down the water supply for several days leaving more than 400,000 people in Toledo, Ohio area without the ability to use water from their tap. This is an epic failure to deliver a basic public service: clean, reliable public water. The response has largely focused on better detection so the pollutants can be treated at the drinking water plant and far less on stopping phosphorus from coming off farm fields and out of farm drain tiles. Solutions should focus on the problems at their source rather than at the end of the pipe. For agriculture, an end of pipe solution is not feasible because the "end of the pipe" is literally at the edge of every field that has drain tiles and there are tens of thousands of fields in the region. Where there is no drain tile, there is no pipe. Containing excess nutrients on the land where they are applied or generated is the optimal solution. Implementing this solution may mean that farmers pass on costs (of protecting water sources to consumers). This true cost of agricultural production has previously been falsely ignored (an externality in economic terms) resulting in polluted waters. There is a need to educate people about the real value of protecting and providing clean fresh water, which would still be fraction of the daily coffee or mobile phone bill.

The Toledo crisis was followed by the Flint water crisis: another epic failure of government at all levels to provide safe, reliable drinking water. Corruption, cronyism, racism, ineptitude and incompetence combined in a veritable perfect storm—poisoning children and depriving an entire population of basic access to clean water. For these incidents to have occurred in a populated area in North America is an embarrassment and a tragedy.

In the 21st century, Great Lakes managers and policy makers can point to having "restored" historically contaminated hot spots around the lakes that had been identified nearly half century ago, but these are relatively small gains against the daily onslaught of pollution from unregulated sources, land conversion and alteration of natural water courses that undermine the basic ecosystem functions and services that allow all other ecosystem services to flourish.

I used to answer the question of how the lakes were doing by responding with a question of "which one"? Or, "in what way?" Even though Lake Superior show signs, for the most part, of a healthy ecosystem mainly because fewer people use it and live around it, and thus are able to harm it. Lake Superior is also the largest of the five lakes by volume—the other four could be dumped into Superior's basin and not fill it—so the volume builds in resiliency. Elsewhere, ecosystem services are compromised or severely degraded by climate change. For example, Great Lakes Fisheries require significant interventions and investments to keep the fishing industry afloat amidst food webs that are a shadow of their historic structure.[17]

New Paradigms: Old Paradigms Revisited

Books have been written about environmental problems in the Great Lakes region. The name "Great Lakes" reflects the large lakes themselves—those vast inland seas. Notwithstanding declining fisheries and disrupted food web dynamics in the open lakes, this chapter focuses on the water-energy-food systems as they play out on the land—in cities, towns and villages—around the lakes themselves, and also in those waters close to shore. To this end, this chapter is more about the energy-water-food nexus in Great Lakes region and how it impacts water resources that flow through the watersheds and that serve as the lifeblood for the tributaries that feed the Great Lakes.

Great Lakes advocates claim significant progress in restoring legacy contaminants protecting some critical resources through the U.S. Great Lakes Restoration Initiative (GLRI)—a multi-year, multi-billion-dollar effort underway since 2010. There is reason to applaud the collective action to remedy environmental ills that have plagued the region for nearly a half century. Unfortunately, GLRI, has been mostly backward looking—cleaning up problems that were created in 20th century—and is based on the same policy framework and cultural paradigm that treats water, air and land as distinct components of the environment that need to be "managed" instead of treating them as integrated systems. By way of example, more than US$4 million GLRI dollars[18] have been spent putting more money into "incentives" for agricultural conservation, with virtually no political will to take the difficult stand of treating drain tiles for what they are: pipes; or, in environmental terms, "point source" discharges. Drain tiles coming off farmland are not substantially different from the pipes from steel and pulp and paper mills, that fueled much of the Great Lakes economy in throughout the 20th century. Drain tiles carry pesticides, herbicides, excess nutrients, sediments from farmland directly into nearby "ditches" or drains, many of which

---

[17]   Egan, D. 2017. *The Death and Life of the Great Lakes.* W.W. Norton & Company. New York, New York.

[18]   Great Lakes Commission, https://www.glc.org/work/reap. Accessed 02/28/2019.

used to be called "streams." Pesticides and herbicides disrupt the bio-chemistry of organisms living in those drains and streams. Excess nutrients cause nuisance and/or toxic algae blooms. Sediment sounds benign enough, but any fishery biologist or stream ecologist will tell you that sediment degrades stream habitats; it's literally a killer.

A paradigm shift is needed that recognizes the interconnectedness of air, water, and land. Systems thinking, and the basic concept of "ecosystem integrity" was briefly popular in the 1970s and 80s. Indeed, the Great Lakes Water Quality Agreement was amended in 1978 to give greater weight to a systems or "ecosystem approach",[19] but eventually the ecosystem approach was largely subsumed by the dominant paradigm of putting the ecosystem into buckets and managing those buckets. This fate is not unique to Great Lakes region. The 2012 protocol that amended the Great Lakes Water Quality Agreement has 10 Annexes to address different challenges facing the Great Lakes. For example, Annex 4 focuses on addressing excess nutrients and Annex 6 focuses on invasive species. The issue-specific annexes that could forsake the needs of the system for the benefit of a singular elements, are counterbalanced other annexes that offer a more systems approach, particularly Annex 10 which focuses on science and Annex 2 which calls for integrating elements of the other topical annexes to improve water quality on a lake by lake basis.

The Great Lakes Water Quality Agreement continues to be the policy backbone of transboundary water governance for the Great Lakes. As a binational executive agreement, however, its power comes when the individual governments that signed the Agreement—the United States and Canada—implement laws that reflect the principles and objectives in that Agreement. The record is uneven here, and reflects in part the nature of slow acting (some may say stable) institutions and influenced by the tides, or perhaps seiches?, of national, and increasingly global economic and political forces.

Transboundary governance for water has the support of the Great Lakes *Water Quality Agreement*, but there is no similar binational agreement, policy or law compelling bilateral action when it comes to more sustainable agricultural or energy practices. Great Lakes water is treasured in the abstract—the idea of these vast bodies of water where people at the water's edge cannot see land on the other side and that support multi-billion fishing and recreational boating industries, as well as a lackluster seasonal commercial shipping industry. However, there is a disconnect between these vast bodies of water and how people interact with the lakes. In reality, the majority of people do not interact with the vast open waters of the lakes, but rather in a narrow band of water near the shore and in tributary rivers that drain into the lakes. That the Great Lakes region's identity is around the *lakes*, and not it's coast, or the land, rivers and streams that feed those lakes can present a challenge for protecting and properly managing the areas humans impact the most. Like the polices that look at environment, energy and agriculture as distinct systems, the region's identity has, to some extent, also been truncated. Far more money is invested in protecting fish in

---

[19]   https://binational.net/glwqa-aqegl/ and Protocol Amending the Agreement Between Canada and the United States of America on Great Lakes Water Quality, 1978, as Amended on October 16, 1983, and on November 18, 1987. Signed September 7, 2012, entered into force February 12, 2013.

the open lakes than protecting headwater streams, tributaries and coastal ecosystems. To be sure, managing the lakes where there are only two ownership interests—the governments of the U.S. and Canada, is far easier than managing rivers and coasts which have hundreds if not thousands, of ownership interests, business investors, and management authorities.

Just because it is harder, should not deter us from correcting our course. A new governance paradigm is needed that re-embraces the ecosystem approach: that considers the coasts, river, tributaries and the land that drains into them as part of the Great Lakes. This can be done by leaders of all sectors in the region. The integrating Annexes in the 2012 protocol to the Great Lakes Water Quality Agreement (i.e., Annex 2, 9, and 10) offer some hope, but the region would be wise to foster a systems approach across all societal sectors. Research institutions and grass roots organizations are a good place to start by clearly articulating the value proposition (what is to be lost or gained). The concept of a nutrient utility[20] reflects systems thinking but likely has a long way to go before it can be integrated into the well-entrenched and fractured water utility paradigm. Citizen groups can amplify this message by putting pressure on their elected and appointed officials and the institutions that employ them. Philanthropy can play a greater leadership role in investing in ideas to test innovative approaches that reflect systems thinking. Emerging efforts by foundations to enable sub-lake regional efforts that marry green and gray infrastructure are a promising example. A well-articulated value proposition should get the attention of businesses and residents who rely on the Great Lakes ecosystem services that provide ample clean fresh water, farmland and forests, and other natural resources. This re-emphasis on ecosystem approach can be home grown–led and carried through by leaders and residents across the Great Lakes region.

Integrating energy and agriculture with water is a heavier lift. Though leaders in the Great Lakes region can begin and continue to advance the discussion, but the forces that drive energy and agriculture are much bigger than the Great Lakes region and will require engaging in dialogues at national levels at a minimum. On the U.S. side, the Farm Bill will continue to drive agricultural policy, but the tensions between commodity price supports, conservation titles that pay farmers to do the right thing can be exposed creating space for new policies that are more just and sustainable. Energy policy is a third-wheel in Great Lakes policy circles. Although there are contradictions, the Farm Bill at least, reflects a national policy on agriculture. The U.S. has no such cohesive energy policy. Energy markets are much more opaque than agriculture; a prime example is the silent but tremendously powerful role of regional transmission organizations. These institutions determine which energy gets to the market and when, but are not well understood by most people who think about transboundary governance in the Great Lakes region. Consequently, bringing energy considerations into an energy-water-food paradigm is likely to be even more difficult.

Research institutions like the National Science Foundation in the U.S. are already showing some promise to consider the food-energy-water nexus and implications of

---

[20]   US Water Alliance, NACWA, and WEF. 2017. Addressing Nutrient Pollution in Our Nation's Waters: The Role of a Statewide Utility.

national and regional governance.[21] Some leaders are suggesting that international and national political demands rely less on global trade and immigration coupled with growth in consumer demand for products more locally-produced goods compelling a fresh look at the food-energy-water nexus. Deglobalization is not likely to happen in the short term, but these forces may enable a more thoughtful and integrated look at how regional ecosystems support regional economies and help better integrate our food, energy and water systems toward a more sustainable future.

Conclusion

This chapter highlights connections between the food, energy and water nexus in the context of the
Great Lakes basin of North America. The complex institutional frameworks at national, sub-national and local levels of government designed to improve efficiencies have had disastrous unintended consequences for ecosystems and the services they provide upon which human society and economic activities rely. Linkages among food, energy and water are starting to get more attention in academic and non-profit organizational circles, but government is slow to reflect this growing body of knowledge and alternative governance forms are not forthcoming quickly enough to solve these wicked problems. Entrenched interests in current institutional and policy structures, particularly among the corporate business community, will be tremendously difficult to change unless such changes can be done without significant disruption to those markets. A new paradigm shift in government and governance that incorporates values can help address some of these policy and market failures.

---

[21] National Science Foundation Innovations at the Nexus of Food, Energy and Water Systems (INFEWS) Program. See https://nsf.gov/funding/pgm_summ.jsp?pims_id=505241.

Chapter-8

Linking the Water-Food-Energy Nexus to Sanitation

Will it Save and Improve Lives?

Sara Gabrielsson,[1,*] *Jamie Myers*[2] *and Vasna Ramasar*[1]

INTRODUCTION

The world is facing increasing pressures on both ecological and human environments. Climate change, rapid urbanization, environmental degradation coupled with economic crises create changing conditions which mean that the world cannot continue with business as usual. With this as a backdrop meeting the ambitious targets set out in the Sustainable Development Goals (SDGs) demand that we reduce long-term dependency on non-renewable resources through the adoption of innovative and adaptive systems promoting recycling and reuse (Cross and Coombes, 2013). Sanitation being one of the least prioritized areas on the global development agenda in the past, due to its high capital investment costs and inherent complexity in behaviour change, technology adoption and implementation, has recently been given more attention for its multiple benefits to human development, when linked to other productive sectors such as energy and agriculture (Drechsel et al., 2011). Nevertheless, global sanitation statistics make depressing reading. The Joint Monitoring programme, a joint initiative between the World Health Organisation and UNICEF, reported that in 2015 2.3 billion people lacked even basic sanitation of which 892 million people practise open defecation (UNICEF/WHO, 2017). The

---

[1] Lund University Centre for Sustainability Studies (LUCSUS), Sweden.
[2] Institute of Development Studies (IDS), United Kingdom.
* Corresponding author: sara.gabrielsson@lucsus.lu.se

impacts of this is direct and long-lasting. Diarrhoea alone is responsible for an estimated 21% of under-five mortality in the global south—2.5 million deaths per year and over 4% of the world's disease burden (J-PAL, 2012).

Not only does open defecation and a lack of safe excreta management cause disease but it also leads to water resources pollution and damages ecosystems and the services they provide (Andersson et al., 2016). The associated deaths and diseases caused by inadequate sanitation also has direct impacts on a country's economy due to demands on the medical system, loss to the labour force and delayed entry into the labour force due to poor performance during schooling (UNICEF/WHO, 2014). In Tanzania, for example, it has been estimated that inadequate sanitation costs US$206 million each year, equivalent to 1% of Tanzanian gross domestic product (GDP) and US$5 per person. The economic losses are directly related to loss of time for people having to find places to defecate, premature death, productivity losses whilst sick and money spent on health care (URT, 2012).

So how can we achieve access to adequate and equitable sanitation and hygiene for all, end open defecation, paying special attention to the needs of women and girls and those in vulnerable situations, while simultaneously recover the resources found in the human waste we produce without contaminating the environment? Is a nexus approach to view and do sanitation a viable way? In this chapter, we explore the usefulness of productive sanitation to ensure that the sanitation demands and needs of the most vulnerable people on the planet, those who are currently reliant on unimproved sanitation facilities and those who practice open defecation, are met and sustained.

Sanitation Status and Health

Exposure to faeces increases the risk of Faecally-Transmitted Infections (FTIs), a term used to cover faecal-oral, parasites which enter through the skin and "water-borne" diseases (Chambers and Medeazza, 2014). Ingesting faeces can lead to diarrhoea and environmental enteropathy, which is estimated to affect almost all children in developing countries, causing chronic child undernutrition and stunting (Humphrey, 2009). Soil and water contaminated with faeces can transmit protozoa and helminth infections which cause poor appetite, nutritional deficiencies and anemia all of which exacerbate malnutrition (Chase and Ngure, 2016).

In order to break the cycle of disease all members of a community must use an improved toilet consistently so contamination and exposure to faecal matter is no longer a risk. The consistent usage of toilets by all household members who have access to toilets is emerging as a major and growing problem, especially in India (Chambers and Myers, 2016). Studies in Lao PDR and Vietnam have shown that children are still at risk of stunting even when families have an improved latrine if a small proportion of households continue to practise open defecation or are reliant on unimproved facilities (Quattri and Smets, 2014; WSP, 2014). In order for maximum benefit to water sources, health, nutrition, education, etc., everyone must use an improved latrine. Even with improved sanitation there is the secondary challenge of how to safely contain and manage faecal sludge. Bad pit latrine management and unhygienic pit emptying processes pose serious health and environmental risks (Evans et al., 2015).

Sanitation and the Linkages to the Water-Energy-Food Nexus

Sanitation is affected by and affects water, energy and food interacting with the nexus at all points. The connection between water and sanitation is generally accepted, however, the complexity of issues and the repercussions for energy and food security both directly and indirectly are less often acknowledged. Given the large sanitation challenges faced by billions of people, the potential for water-food-energy security to be achieved and the SDGs to be met is jeopardized by this sanitation backlog. By looking at the system holistically, which addresses the linkages between sanitation, and water-food-energy nexus multiple benefits across several fronts can be achieved (Andersson et al., 2016). In this section we describe some of the inter-linkages between sanitation and water, energy and food.

In many parts of the world, especially cities of the global south, water demand for productive, domestic consumption and hygiene purposes are higher than the delivered safe supply. This leads to people obtaining water from open sources that are at risk of contamination. Many people, especially in urban informal settlements, where improved sanitation is the lowest, get their water from shallow wells or open water bodies. The shallowness of these wells make them especially vulnerable to water pollution and the major cause of the pollution can be linked to unimproved sanitation facilities and improper disposal of faecal sludge, particularly linked to the common practice of opening up toilets to flush the sludge away during heavy precipitation events (Andersson et al., 2016). This can lead to the contamination of water resources with both faeces and urine. Water contaminated with faeces increases the risk of FTIs while urine can cause the eutrophication of fresh water resources. During times of flooding, contaminated water can spread diseases (Kiunsi, 2013) and also serve as breeding grounds for other parasites. Risks to groundwater from improper sanitation practices do exist but these have been found to be low (Sugden, 2016; Cave and Kolsky, 1999). Improper sanitation practices therefore are a risk to water quality and quantity of safe water supply.

However, the situation is further complicated as the use of improved sanitation and hygiene facilities often increases the demand for water for operation, use and cleaning. In addition, hand-washing, safe menstrual hygiene management, cleaning of latrines and anal cleaning all increase water demand. This is especially so in contexts where access to cleaning products such as toilet paper and sanitary pads are not easily available or are expensive to obtain.

Sanitation may affect energy use in several ways and at different scales from the body to the household energy use and national infrastructure investment. At the scale of the human body, energy to do physical labour is compromised when water-borne diseases affect nutritional uptake. Human energy is also wasted if household members, predominantly women, have to spend longer travelling to collect water if closer sources are contaminated. Toilets, which consume large amounts of water will also have the same effect for some households. Households also face a trade-off between paying for energy needs and paying for improved sanitation, which often leads to poor investment in improved sanitation. Similarly, on a national scale, government investment in energy infrastructure may take precedent over investment in sanitation.

Developing further the relationship between sanitation and human energy, the connection to food and nutrition becomes visible. Most directly, FTIs are spread through faecal-oral infections often associated with food hygiene. In addition, open defecation sites are often areas that could be used to grow fruit and vegetables. When FTIs do occur they affect adsorption, appetite and nutritional deficiencies making food intake less effective (See BOX 1). Medical expenses can take money away from investment in farming equipment, higher quality seeds and crops, fertilizers and food. Sickness caused by poor sanitation can in turn reduce a person's capacity to work and engage in agricultural production (Krishna, 2010).

There are thus multiple pathways that connect sanitation to the water-energy-food nexus. Many of the elements associated with poor sanitation can be reduced thereby improving sustainable development. However, a word of caution is needed as improved sanitation can lead to greater demand for water and energy. Flush toilets for example, need large amounts of water, diverting water away from drinking, other household demands and agricultural use as well as increasing the burden on women and girls who are usually tasked with collecting it. However, many sustainable/ productive sanitation options are promoted as using less water compared to water based systems enjoyed by almost everyone in the developed world. In addition, the safe reuse of urine and faeces can act as a fertilizer, soil conditioner and humus replenisher. Fish can be farmed in ponds where effluent or sludge is dumped. Fish remove the nutrients from the wastewater and provide a cheap protein source (Tilley et al., 2014). There are thus opportunities to improve the water-energy-food nexus by addressing sanitation.

BOX 1: WASH and Nutrition

The relationship between Water, Sanitation and Hygiene (WASH) and nutrition is much stronger than originally thought and the WASH sector is now being recognized as key for maximizing nutritional impact (Chase and Ngure, 2016). It has been estimated that universal coverage of the most effective traditional nutritional programmes in high burden countries would only reduce stunting by 20% globally (Ibid). Nutrition programmes dominantly focus on food intake and assuring enough reliable good quality is continuously available (Chambers and von Medeazza, 2014; Humphrey, 2009). However, FTIs affect adsorption, appetite and nutritional deficiencies making food intake less effective.

Environmental Enteric Dysfunction (EED): Evidence now points to poor sanitation and the ingestion of faecal bacteria as major cause of EED (Humphrey, 2009) a subclinical condition that leads to damage of the small intestine reducing absorptive capacity (Crane et al., 2014).

Is has been estimated that 60% more food will be produced by 2050 to feed the worlds growing population (FAO, 2014). If children continue to live in environments where they are continuously exposed to faeces this food will not be as effective as it could be.

Productive Sanitation—Harnessing Synergies within the Water-Energy-Food Nexus

Since the turn of the millennia the interest in reuse and recycling options as a way to tackle the sanitation challenge as well as protecting natural resources has been given more attention. The Bellagio principles adopted in 2000 set the departure point for what sustainable sanitation systems ought to include. In short sustainable sanitation systems refers to systems that; promotes health and hygiene effectively; is financially and economically viable; is socially acceptable and institutionally appropriate; use appropriate technology, including operations and maintenance while simultaneously protecting the environment and natural resources (SuSanA, 2016). Productive sanitation is the term used for the variety of sanitation innovations that make productive use of the nutrient, organic, matter, water and energy content of human excreta and wastewater for agricultural production, aquaculture and/or energy use (SuSanA, 2016). The term's increased use in academic circles comes at a time of dwindling availability and access to fertile land, freshwater and forests and it has been proposed that with 'the right sanitation investments made now can help to meet rising demands for clean water, food and energy' (Davis et al., 2014). By drawing on knowledge developed in a number of sectors it is believed that the nexus approach can be used to attract interest and capacity from different sectors, individuals, local governments and public-private partnerships to harness synergies to contribute to sustainable food-, water- and energy security in the future (ibid).

In the water sector, greywater from wastewater plants can be used for irrigation when treatment has been completed, improving irrigation possibilities in areas currently with a scarce water supply. In Egypt, Brazil and Namibia it has already been proven that reusing wastewater for irrigation can be a success (Andersson et al., 2016). In the energy sector, faecal sludge can be turned into biogas or bio-charcoal or pellets that can supplement fuel wood and charcoal and thus be considered a mitigation strategy that reduces deforestation rates. At the same time, revenue from bio-charcoal production can promote both increased provision of on-site sanitation technologies and collection of faecal sludge, reducing the health, environmental and risk implications of deficient sewerage systems, as well as serve as a source of income for households (Drechsel et al., 2015). In the agricultural sector both urine and treated human faecal matter can be turned into directly applied fertilizers or provide feed for chickens, larvae or soldier flies, that later can be applied to crops (Andersson et al., 2016). This will reduce the dependency on chemical fertilizers, which currently in many developing countries are transported from far away places or imported at a very high cost to end-users. Moreover, many of these technologies are locally produced and ecologically friendly, thus come at a lower cost per unit. But questions have been raised about whether or not technologies that are more ecologically friendly are less effective at reducing risks to health, especially reducing FTIs (Smet and Sudgen, 2006; Satterthwaite et al., 2015). From a public health perspective faeces is the real danger and the most important thing to contain and treat. Those using a nexus approach to attract interest and funds to implement productive sanitation technologies must therefore make sure the human health aspect is not ignored in pursuit of reuse and recycling.

Moreover, there are significant differences in the way sanitation can contribute to the WEF nexus in an urban and a rural developing country setting, and each geographical space has its own challenges and opportunities, which we will discuss in more detail next.

Productive Sanitation in Urban Areas

Scaling-up efforts to reach the SDG on water and sanitation for all, especially in the fast growing cities of Sub-Saharan Africa (SSA) and Asia, will require more attention to the productive linkages between sanitation, energy and the agricultural sector. The chief reason why the cities of Sub-Saharan Africa, in particular, begs our attention, is because its population has more than doubled from 1990 to 2010, from 145 million to 337 million, increasing the demand for water, food and energy to meet basic livelihood needs (UNEP/UN-HABITAT, 2010). Two thirds of these people today live in unplanned and underserved settlements, which are growing at a rate of four to five million people per year. Here access to and quality of sanitation facilities are exceptionally poor and 80% of the population rely on on-site technologies, including traditional and improved pit latrines, unsewered public ablution blocks, septic tanks, aqua privies, and dry toilets (UNICEF/WHO, 2014). Pit latrines are by far the most commonly used on-site facility across the globe, in both urban and rural areas. These latrines are usually poorly built and maintained and are therefore at continuous risk of contaminating the surrounding environment and posing risks human health (Chowdhry and Koné, 2012). Management of the Faecal Sludge (FS), is by and large, ignored by the public authorities across the global south. A majority of cities therefore rely on informal services for excreta disposal and the business of emptying and transporting FS is dominated by private entrepreneurs, who, due to lack of options, dump large amounts of untreated FS into natural or man-made drainage ways (UNEP/UN-Habitat, 2010). Indeed, it is estimated that across the globe today 2.7 billion users of on-site sanitation systems generate faecal sludge that goes largely untreated, and that number is expected to grow to 5 billion by 2030 (Strande et al., 2014). The emphasis on Faecal Sludge Management (FSM) to deal with the increasing loads of faecal sludge produced in urban areas is therefore gaining traction in the sanitation sector, not only for its key role in improving the containment, collection, transportation and treatment of FS, but also as a way forward to turn human waste into value derived products (Strande et al., 2014; Dodane et al., 2012; Diener et al., 2014; Muspratt et al., 2014; Andersson et al., 2016).

Barriers to improved FSM in urban areas

A number of key overarching barriers to improve overall FSM and provide an enabling environment to turn excreta into value has been identified.

First—*policies*—in most countries of the global south a range of ministries has policies and acts that relate to sanitation and these policies and legislation have broad sanitation objectives; however, few mention faecal sludge management and its aspects specifically (Bahri, 2012). In SSA, in particular, sanitation regulation is complicated by overlapping lines of authority between different government departments, and in most instances these high-level policies and legislation do not link sufficiently to by-laws and implementation at the local level, whom in practice, through the local authority (under different names), is ultimately responsible for faecal sludge management (WRC, 2015). Capacity for implementation and law enforcement at local levels are therefore limited and weak. The responsibility for the management of FS thus falls on the users of the system, predominately the urban poor, who largely lack capacity to claim their rights to reliable sanitation services and the awareness of the potential value that productive sanitation may have on their lives (ibid). A direct implication of the lack of FSM policies and regulations is that private manual FS collectors are still not recognized by the law, making their job particularly hazardous and difficult to perform (Chowdhry and Koné, 2012).

Second—*financial means*—national budgets allocate significantly more funding to urban water than to urban sanitation and the focus of sanitation budgets is mainly on the provision of sanitation facilities, and less on operation and maintenance, including FSM (Strande et al., 2014). Donor funding and other aid to sanitation is also still focused on infrastructure development and health and hygiene education (ibid). As a result, local authorities do not have enough budgeted funds for faecal sludge management (WRC, 2015). Consequently the main economic burden for the construction and emptying of pit latrines in cities are currently put directly on households. Households who predominately live in informal settlements with generally low ability and/or willingness to pay for such infrastructure and services, due to competing livelihood demands (McGranahan and Mitlin, 2016). In Accra, Ghana, for example, the cost to poor households for emptying their latrine is 10 times more than the percentage of household income that is considered to be equitable for sanitation services (Boot and Scott, 2009). To reduce the costs associated with pit emptying many households resort to on-site disposal (with or without treatment) by either closing a full pit and digging a new one nearby or flushing out FS from the pit during heavy down pours into the surrounding environment (Strande et al., 2014). Shifting the financial burdens of emptying services away from urban poor households is therefore seen as a must to improve performance of FSM.

Third—*spatial context*—informal settlements are naturally unplanned and densely populated and continuosly expanding. Containment of FS is therefore even more important in an urban context than in a rural one due to limited space, high number of potential FS contact areas and the multiple disease exposure routes. The spatial context may also prove a barrier for reliable FS collection. The narrow walk ways to access single households and public ablution blocks make both mechanical and manual pit emptying a time-consuming and costly endeavour. Studies indicate that the majority of these businesses are currently single truck operations, and they have much lower profit margins than businesses with bigger fleets (WRC, 2015). But currently no special financing schemes are available to these private entrepreneurs who want to expand their fleet and business sizes to be more financially viable (Chowdhry and Koné, 2012). Moreover the increasing presence of solid waste (specifically plastics) inside the pits is also filling up pits faster making collection more challenging and reducing the actual amount of FS collected at each house. While education programmes might improve the situation, unless the responsible authority removes solid waste regularly from informal settlements, it is unlikely that this problem will be resolved (WRC, 2015). Space also matters when it comes to treatment and reuse of FS. Traditional open-air drying beds, require land to enable dewatering of FS, land that may be prohibitive in many urban areas (Muspratt et al., 2014). Optimizing fast and effective drying techniques that require less land is therefore a research area of vital importance to improve FSM and resource recovery.

Fourth—*social stigma*—handling of excreta is still not socially accepted by the majority of people in Africa and South Asia. In Sub-Saharan Africa FS collection and transportation is therefore largely done by small groups of manual collectors in the middle of the night for fear of being recognized and arrested by the authorities, but also as a strategy to reduce harassment from people who think their work is a taboo (Chowdhry and Koné, 2012). Systematic studies of existing working conditions of these so called 'frog men' are largely missing, but there is little doubt that the social stigma attached to handling excreta increases the occupational hazards that comes with the job, as the service they provide is not valued and paid enough to ensure that the workers have protective gear to safeguard them from direct contact with pathogens (Kiunsi, 2013). Consequently, the social stigma also spills over into an initial stark scepticism about any product derived from excreta. Particularly products that are intended to have direct contact with food or still look like or resemble human waste.

Fifth, *existing treatment facilities*—in many countries across the global south only a few faecal sludge treatment plants and waste-water treatment plants exist, and many of these are dysfunctional, overloaded and located far away from the city (Strande et al., 2014). For example, in Jakarta, "less than three per cent of the 1.3 million cubic meters (enough to fill more than 500 Olympic swimming pools) of sewage generated each day reaches a treatment plant—there is only the capacity to process 15 swimming pools' worth" (Corcoran et al., 2010). Moreover while FS collectors are required to transport the FS to existing treatment facilities, many still save time, money and fuel by dumping the sludge illegally by the roadside or in open spaces in the cities, creating a health hazard for everyone (Chowdhry and Kone, 2012). Investments in new FS facilities and repair of old ones is a must to capture the unquantified FS now dumped throughout the cities of the global south.

Opportunities of productive sanitation in urban areas through improved FSM

While there are critical obstacles to improve FSM and enable resource recovery from FS there are also burgeoning opportunities if the barriers mentioned are managed.

First—*FS is readily available and contain high caloric values*—as such the technical and financial viability for resource recovery are vast and identification of products in use today that could be directly replaced by FS in a global south context are many (e.g., solid fuel, fertilizer), or derived from faecal sludge (e.g., animal feed, biogas) (Diener et al., 2014) or as part of building materials and use as a soil conditioner (Muspratt et al., 2014).

Second—*shifting financial burdens for pit emptying away from the urban poor*—by involving private actors in resource recovery from collected FS it may be possible to alter the financial flow of the sanitation service chain, and offset some of the disposal costs and thus reduce the direct financial costs to households of pit emptying

(Koné, 2010; Diener et al., 2014; Muspratt et al., 2014). In turn this may serve as one incentive to change current on-site disposal practices and reduce the risks of groundwater contamination and linked health and well-being impacts, thereby also contributing to an increase in overall sanitation service delivery.

Third—*FSM help reduce overall costs for national governments*—it has already been demonstrated that FSM technologies have overall annualized capital and operating costs that are five times less expensive than conventional sewer based solutions (Dodane et al., 2012). Implementation of sound FSM system could thus divert political attention, and financial resources, away from the idea of the need for capital intensive conventional sewerage systems and focus it on more financially and contextually more viable sanitation solutions that may reach the most in need of sanitation services.

Four—*Creation of jobs and income opportunities for locals*—several new private initiatives have proven that productive sanitation can indeed contribute to creating jobs that previously did not exist. For instance, *Sanergy* in Kenya has created a franchise business-model creating local jobs in informal settlements for anyone interested in running a 'Fresh Life' toilet. These so called 'Fresh Life Operators' (FLO) are local residents who purchase and operate urine diversion toilets with ongoing operational and marketing support, and a daily waste collection service facilitated by the mother company Sanergy. The responsibility of the FLO is then to generate local demand and ensure that the Fresh Life toilet is kept clean. To date a total of 732 FLOs are active in Nairobi, with 37,238 daily uses from community members and a total of 8,504 metric tons of human waste safely removed from the community and safely treated (Sanergy, 2016).

Five—*Builds social cohesion and mutual accountability within communities*—Harnessing human waste and turning into value also has the promise of building stronger communities. In informal settlements across Kenya, the local NGO *Umande Trust* has proven that entrusting local community groups to run public bio-centres that offers paying community members access to a shower, clean toilet and cooking facilities, it is possible to build social cohesion and mutual accountability even in the poorest and most marginalized communities on the planet. Here community groups are responsible for the management of a dussin public bio-centres that turns human waste into bio-gas and provides local groups with the financial means to sustain their lives. The bio-centres not only provide a vital sanitation service to the public by collecting, containing and treating human waste, it also provides the accountability needed from within to continue to manage the centres and the financial means to keep them going in the long term (Umande Trust, 2016).

Six—*Boosting agricultural and industrial development*—tremendous opportunities exist to turn human waste into products for use in both the agricultural and energy sectors. Urine is the most directly applied product for use as a fertilizer. *Sanergy* is already collecting urine from their Fresh Life toilets for use in agricultural production. Simultenously they are also treating and turning faecal sludge into high-quality manure to be sold and used by farmers outside Nairobi. In Accra, Ghana the private company *Slamson Ghana Ltd.* is doing the same, but on a much larger scale, dewatering faecal sludge from about 80 trucks per day and turning it into both manure and bio-charcoal. *Sanivation* in Kenya is also producing bio-charcoal, but using a different business model, where customers pay a small fee for having feacal sludge collected from their individual container based toilets. In Rwanda the social enterprise *Pivot Works* is taking stock of the energy needs in the growing industrial sector in the country, particularly targeting the cement industry, which is the most energy intensive industry globally. By collecting faecal sludge and turning it into bio-pellets they have created a renewable product that never goes out of season, is cheaper than other biomass fuels and produced locally. Moreover, compared to other biomass fuels their fuel contains 20% more energy, 10% less moisture and emitts less carbon dioxide (Pivot Works, 2016). Studies from other countries also indicate the promise of using faecal sludge as a dry combustion fuel in industry (Boesch and Hellweg, 2010; WBCSD, 2005). And laboratory and pilot scale research conducted in both Dakar and Kampala has demonstrated its technical viability (Muspratt et al., 2014). Certainly the main potential market for sludge as fuel has been identified as the industrial sector; since industries have both the fuel demand and ready willingness to use the product. Depending on the city, country and type of industry, a variety of fuel sources could be supplemented or replaced with faecal sludge (Diener et al., 2014).

Beyond bio-charcoal and the direct use of human waste as manure and fertilizers there are also opportunities of using faecal sludge as a medium for rearing insect larvae for protein in animal feed. To date, there are no full scale implementations, however at the laboratory scale the use of faecal sludge as a feed source for Black Soldier Fly (BSF) larvae, Hermetia illucens, has been successfully demonstrated by Nguyen (2010). Similarly, in South Africa the company Agriprotein is operating a treatment plant processing waste with insect larvae for the production of chicken and fish feed (van Huis et al., 2013). Given that the global price for fish meal has tripled from 2005 to 2013, and is expected to remain high due to declining wild fish stocks and the on-going increase of aquaculture, the revenue potential from larvae is very attractive (Naylor et al., 2009).

Seven—*Scaling up sanitation service delivery and contributing to climate resilence and mitigation*—The sheer amount of faecal sludge produced in cities across the global south and the fact that the market to date is highly under-leveraged also promises high potential for scalability. Studies show that the calorific value of FS is not particularly sensitive to the source (i.e., pit versus septic) or city, suggesting that the use of FS as fuel or for agricultural uses could be easily transferred across cities and countries (Muspratt et al., 2014; Andersson et al., 2016). Hence, if valuable faecal sludge-derived products were to be generated, the volumes of raw sludge that are properly collected and managed in cities would most likely greatly increase as a result of the financial market incentive (Diener et al., 2014). But it is important to note that this is contingent on reversing the sanitation value chain so that the financial burdens for pit emptying is shifted away from the urban poor. Also it is imperative to consider the local context and market when selecting and designing treatment and end-use products, as cultural appropriateness and markets vary significantly among locations. If these factors are taken into consideration, reuse of human faecal sludge can offer an affordable alternative with benefits for both energy and food security and sanitation challenges, as well as contribute to climate mitigation and adaptation in urban areas of the global south (Drechsel et al., 2015).

Productive Sanitation in Rural Areas

Approximately 70% of those without improved sanitation facilities and 90% who practise open defecation live in rural areas (UNICEF/WHO, 2017). Up to now the focus has been on increasing the proportion of people with access to improved sanitation and safe containment. The safe management of pits and focusing further down the sanitation chain has been confined to urban environments where high population densities mean the poor disposal of sludge is more of a concern for human health. However, we will need to consider the safe management of excreta permanently in rural environments to ensure health and environmental benefits. This will become even more important as households climb the sanitation ladder and begin to invest more in infrastructure making it more difficult to simply dig another pit. Alongside the benefits to water-energy-food, resource recovery, if done correctly, is one way to ensure that human waste is safely managed. There are a number of opportunities to promote this more holistic system. However, there are also a number of challenges.

Barriers to resource recovery in rural areas

In rural areas there are different implications for households that use productive sanitation. Unlike in urban areas were FSM services may exist, in rural areas this is not a viable option. Households will have to manage the content of the pits themselves. Social acceptance of containment, handling and reuse are therefore more of a concern.

Reuse and recycling options do mean that households are able to do this without the need for external services. However, households can be unhappy about this extra burden placed on them and often would rather have toilets which require no further handling by users (Smet and Sugden, 2006). A study in Indonesia found that despite 82% of respondents believing they would benefit from using an EcoSan Toilets only 50% said they would be willing to process the excreta themselves (WSP, 2010). In Vietnam, a Plan International intervention had been partly subsiding urine diversion double-vaulted latrines. Despite them being popular in Vietnam 78% households with them would still prefer a septic tank and households who were not part of the intervention were waiting until they had enough money for a septic tank (Cole et al., 2008). The handling of faeces and its reuse can also conflict with deeply held cultural beliefs and cultural and religious stigma leading to latrines being unused (Movik and Mehta, 2010). Beliefs are not universal and will be different within and between countries. In areas with less cultural resistance there is potential for resource recovery (ibid).

At the very minimum, sanitation systems must be socially accepted in order to ensure long lasting use by all. The use of excreta for energy production is less controversial than the use in agriculture. However, it has been suggested that household toilets will not generate much methane and the costs involved may not be worth it (Satterthwaite et al., 2015). This consideration is more important in rural areas where there is less money available and lower population density means each household is likely to have their own individual toilet rather than be reliant on community toilet blocks other than in schools and hospitals.

Opportunities for productive sanitation in rural areas

When describing the nexus approach the Stockholm Environmental Institute argues that in order for it to be used in policy and practice it must meet the needs of those managing local resources (Davis et al., 2014). For productive sanitation in rural areas households are those managing the resource. It is therefore essential that any sanitation option meets their needs. This will involve ensuring household participation from the beginning of a process. All members, including the marginalized, must be given the opportunity to join the conversation. If people are unhappy about using a particular technology, do not promote it. For those wanting to use nexus thinking to tackle the sanitation challenge it is important that this thinking is cascaded down to communities and households who will ultimately be the ones making the decision to invest in technologies, use toilets and safely manage excreta.

Though acceptance may be a problem there are certain community level actions that can be taken and information provided to make resource recovery options more appealing. For example, presenting many options at different price ranges with information about the emptying process and what is required from households. These include those which encourage reuse and ones that do not. Households and communities should know all the available practical options and their costs in order for them to make an informed choice. Demonstrating the handling and use can show how easy and non-offensive use is as well as its benefits (Tilley et al., 2014) but do not push for one particular model. Households need appropriate knowledge: where reuse technologies are in use households need to be aware of the periods of time needed for the excreta to be contained before it becomes safe. Researchers found that in an area where double-vault composting latrines had been promoted vaults were being emptied when they were full rather than after the recommended six-month storage time (Cole et al., 2008).

In order to change norms and attitudes to reuse finding technologies that require no direct handling or that transform faeces into something unrecognizable to fresh faeces could be helpful (Drangert, 2004–05). There are a number of options that make the handling and reuse more socially acceptable. Examples include:

- Arbourloos are simple toilets that do not require families to directly handle the pit content. The toilets consist of shallow pits with a temporary or mobile slab and superstructure. When the pit is almost full the slab and superstructure are removed. The pit is then filled with soil and a tree is planted. It can help reforest areas and fruit trees can be grown that can be cultivated. However, it requires a large amount of space and regular reconstruction (Tilley et al., 2014).

- Tiger Toilets are onsite systems that use a vermifilter (worms) to process wet human faeces into dry vermin-compost making it easier to empty the pit and transport to field. The worm's process fresh excrement entering the system daily meaning there is no build-up of waste and no smell. Dry odourless hummus, which is easy to empty and safe to generate at the top of the pits. Small-scale trials in Uganda, Myanmar and India have all shown user feedback to be positive (Furlong, 2016).

Finding relevant community groups to engage and act as gate-keepers will be important. Some countries through previous sanitation drives have seen the establishment of appropriate institutional arrangements at the village level that can be engaged. For example, the Community-Led Total Sanitation (CLTS) approach, used in 50+ countries recommends the formation of sanitation committees (Kar and Chambers, 2008). In Nepal, the Governments National Sanitation and Hygiene Master Plan (2011) clearly spelled out the need of committees at different levels to achieve ODF and led to the creation of Water, Sanitation and Hygiene Coordination Committees WASH-CCs at village, district and regional levels across the whole country. The Community Health Club (CHC) approach sees the establishment of voluntary organizations, CHCs, to tackle a range of health challenges communities may face, including sanitation (Waterkeyn and Cairncross, 2005). It has been used extensively in Zimbabwe, and to a lesser extent in the Democratic Republic of Congo, Guinea Bissau, Rwanda, Sierra Leone, South Africa, Tanzania, Uganda and Vietnam. Both sanitation committees and CHCs are possible village level structures that could be engaged to discuss productive sanitation and nexus thinking.

As mentioned above the health consequences of children growing up in faecally contaminated environments are dire. NGOs, governments and research institutes across the world are grappling with the challenge of making sure people who have toilets consistently use them. This is no easy feat and requires changes in social norms. Everyone must want to use a toilet all the times and expect others to do the same. Full pits and the fear of pits filling up have been identified as a factor associated with non- and partial usage of latrines, predominantly in India (Chambers and Myers, 2016). Excreta management practices should not deter use. Resource recovery must change norms regarding the handling and reuse of excreta.

When is the Nexus Approach to Sanitation a Help or a Hindrance?

It is important to define what is or is not a nexus question (Davis et al., 2014) and assess whether a nexus lens is suitable to ensure equitable access and sustainable use. After looking at the barriers and opportunities to productive sanitation in urban and rural areas it is possible to identify how it can help or hinder efforts.

A help

First and foremost, as has been stated explicitly throughout, the number of those reliant on unimproved sanitation and who practice open defecation is huge and the health consequences dire. By demonstrating the benefits improved sanitation can have, not just on health and dignity, but on water sources and potentially food and energy systems it further highlights the importance of improved total sanitation. The evidence on the linkages and mutual benefits can be used to secure government support, investment and commitment, helping to create the right enabling environment at the international, national and sub-national level for successful and sustainable sanitation programmes.

A nexus approach forces us to take a systems view examining the potential benefits and trade-offs of particular sanitation options. It pushes people to work with colleagues who are not traditional allies. For example, highlighting the links between nutrition and WASH has opened up a dialogue and collaboration between non-traditional partners. Though this does not mean that every sanitation programme should include a nutrition component of vice versa there is now much more of an understanding that mutual benefits do exist.

It is stated in the literature that nexus thinking is more than just a conceptual approach used to analyze linkages between water, energy and food systems. It encourages stakeholder interaction through stakeholder dialogues (FAO, 2014) yet there is no explicit mention of the poorest or most vulnerable (Biggs et al., 2015). Those without access to improved sanitation are the poorest and most vulnerable. Furthermore, due to the nature of the topic households are essential stakeholders in the safe containment and management of sludge in both rural and urban communities. It could be a great tool for the WASH sector both for research and action provided there is space, time and mechanisms in place to cascade nexus-thinking down to the communities. Nexus tools should be highly participatory taking into account peoples preferences and the context which they live in. Discussions need to tackle social norms about toilet use and resource recovery and aim to increase community buy-in. If done correctly there is a higher chance that toilets and FSM systems are operated, maintained and used sustainably.

A hindrance

However, there are also a number of potential traps. Sanitation habits are private and personal and changing behaviour is essential in order to have the maximum health impacts. There is a danger that a systems approach with multiple mutual benefits and trade-offs clouds the fact that the most desired outcome is improved public health. The most pressing advantage of improved sanitation is stopping the spread of FTIs, which stop children not reaching their full potential and dying. From a public health perspective faeces is the real danger and the most important thing to contain and safely manage, however urine is more productive and useful for agricultural production. Furthermore, questions have been raised about whether or not technologies that are more ecologically friendly are less effective at reducing risks to health or even adequately from FTIs (Smet and Sudgen, 2006; Satterthwaite et al., 2015). Those using a nexus approach to facilitate discussion, attract interest and funds must make sure the human health aspect is not ignored in pursuit of energy or agricultural goals.

Finally, those pushing productive sanitation are often able to go home to their flush systems. It should also not be used to determine what technologies the poorest could be using especially when there is a risk that households may be reluctant to use latrines that require them to handle the pit content.

Conclusions

Our work illustrates that there is still a great deal of work ahead in addressing the global sanitation challenge. The problem is a social, technological, environmental, economic and political one and a nexus approach can help to broaden the conversation so that we find areas of mutual benefit that further the mission of safe and productive sanitation. Perhaps most critically, the Sustainable Development Goals articulate that we should leave no one behind. The days of small scale programmes and cherry picking easy communities are passed. Whatever is envisioned needs to be planned on a grand scale at the beginning. This does not however mean rolling out the same for everyone but rather contextualizing and modifying options across different environments and social groups.

In moving forward, more research is needed to understand the potential health impacts of the reuse of human waste. The results of this as well as safe ways of handling and utilizing waste needs to be communicated with people in a manner that is inclusive and respectful of people's social, cultural and emotional hesitancy. In suggesting affordable and scalable technologies for productive sanitation, it is important that we are all cognisant of the fact that many of us promoting such technologies live with the luxury of flush-and-forget systems that are themselves wasteful of water and energy. Communication and engagement of people is critical to the success or failure of efforts for sustainable/productive sanitation. Using arguments that extend benefits across the water-food-energy nexus may be one means of engendering greater support for such systems.

References

Andersson, K., Rosemarin, A., Lamizana, B., Kvarnström, E., McConville, J., Seidu, R., Dickin, S. and Trimmer, C. 2016. Sanitation, Wastewater Management and Sustainability: from Waste Disposal to Resource Recovery. Nairobi and Stockholm: United Nations Environment Programme and Stockholm Environment Institute.

Andersson, K., Dickin, S. and Rosemarin, A. 2016. Towards "sustainable" sanitation: challenges and opportunities in urban areas. *Sustainability* 8(12): 1289.

Bahri, A. 2012. Integrated Urban Water Management TEC Paper 16: Global Water Partnership Technical Committee.

Biggs, E.M., Bruce, E., Boruff, B., Duncan, J.M., Horsley, J., Pauli, N., McNeill, K., Neef, A., Van Ogtrop, F., Curnow, J. and Haworth, B. 2015. Sustainable development and the water–energy–food nexus: A perspective on livelihoods. Environmental Science & Policy 54: 389–397.

Boesch, M.E. and Hellweg, S. 2010. Identifying improvement potentials in cement production with life cycle assessment. *Environ. Sci. Technology* 44(23): 9143–9.

Boot, N.L.D. and Scott, R.E. 2009. Faecal sludge in Accra, Ghana: Problems of urban provision. *Water Science and Technology* 60.3: 623–631.

Cave, B. and Pete Kolsky. 1999. Groundwater, latrines and health. WELL Study 163.

Chambers, R. and von Medeazza, G. 2014. Reframing undernutrition: Faecally-transmitted infections and the 5 As. *IDS Working Paper* 450. Brighton: IDS.

Chambers, R. and Myers, J. 2016. Norms, Knowledge and Usage, Frontiers of CLTS: Innovations and Insights Issue 7, Brighton: IDS.

Chase, C. and Ngure, F. 2016. Multisectoral Approaches to Improving Nutrition: Water, Sanitation and Hygiene. Water and Sanitation Program: Technical Paper. Washington: WB.

Chowdhry, S. and Koné, D. 2012. Business Analysis of Faecal Sludge Management: Emptying and Transportation Services in Africa and Asia. Seattle: The Bill & Melinda Gates Foundation.

Cole, B., Phuc, P.D. and Collett, J. 2008. The sum is greater than the parts: An investigation of Plan in Vietnam's double-vault composting latrine program in northern Vietnam. *In*: WaterAid Australia and International Water Centre (ed.). Sharing Experiences: Sustainable Sanitation in South East Asia and the Pacific.

Crane, Rosie J., Kelsey D.J. Jones and James A. Berkley. 2014. Environmental Enteric Dysfunction—an Overview, CMAM Forum Technical Brief, Collaboration to improve the management of acute malnutrition worldwide, August.

Cross, P. and Coombes, Y. 2013. Sanitation and Hygiene in Africa: Where do We Stand? IWA Publishing.

Corcoran, E., Nellemann, C., Baker, E., Bos, R., Osborn, D. and Savelli, H. (eds.). 2010. Sick Water? The central role of waste-water management in sustainable development. A Rapid Response Assessment. United Nations Environment Programme. GRID-Arendal: UN-HABITAT.

Davis, M., Huber-Lee, A., Hoff, H. and Purkey, D.R. 2014. SEI Research Synthesis: The water-energy-food nexus. Stockholm. Sweden.

Diener, S., Semiyaga, S., Niwagaba, C.B., Muspratt, A.M., Gning, J.B., Mbéguéré, M., Ennin, J.E., Zurbrugg, C. and Strande, L. 2014. A value proposition: Resource recovery from faecal sludge—Can it be the driver for improved sanitation? Resources, Conservation and Recycling, 88: 32–38.

Dodane, P.H., Mbéguéré, M., Sow, O. and Strande, L. 2012. Capital and operating costs of full-scale fecal sludge management and wastewater treatment systems in Dakar, Senegal. Environmental Science & Technology 46(7): 3705–3711.

Drangert, J. 2004–5. Norms and Attitudes Towards Eco-san and Other Sanitation Systems. EcoSanRes Publications Series.

Drechsel, P., Cofie, O.O., Keraita, B., Amoah, P., Evans, A. and Amerasinghe, P. 2011. Recovery and reuse of resources: enhancing urban resilience in low-income countries. Urban Agriculture Magazine 25: 66–69.

Drechsel, Pay, Manzoor Qadir and Dennis Wichelns. 2015. Wastewater: Economic Asset in an Urbanizing World. Wastewater: Economic Asset in an Urbanizing World 1–282.

Evans, B., Fletcher, L.A., Camargo-Valero, M.A., Balasubramanya, S., Rao, C.K., Fernando, S., Ahmed, R., Habib, M.A., Asad, S.M., Rahman, M.M. and Kabir, K.B. 2015. VeSV-Value at the end of the Sanitation Value Chain.

FAO. 2014. The state of food insecurity in the world 2014: strengthening the enabling environment for food security and nutrition. FAO: Rome.

Furlong, C. 2016. Tiger Worms: A win-win solution. *In*: Chambers, R. and Myers, J. (eds.). Norms, Knowledge and Usage. Frontiers of CLTS: Innovations and Insights Issue 7, Brighton: IDS.

Humphrey, J. 2009. Child undernutrition, tropical enteropathy, toilets, and handwashing. The Lancet 374.9694: 1032–1035.

J-PAL. 2012. J-PAL Urban Services Review Paper. Cambridge MA: Abdul Latif Jameel Poverty Action Lab.

Kar, K. and Chambers, R. 2008. Handbook on community-led total sanitation. IDS: Brighton. UK.

Kiunsi, Robert. 2013. The Constraints on Climate Change Adaptation in a City with a Large Development Deficit: The Case of Dar Es Salaam. Environment and Urbanization 25(2): 321–37.

Koné, D. 2010. Making urban excreta and wastewater management contribute to cities' economic development: a paradigm shift. *Water Policy* 12(4): 602–10.

Krishna, A. 2010. One Illness Away: Why People Become Poor and How They Escape Poverty. Oxford: Oxford University Press.

McGranahan, G. and Mitlin, D. 2016. Learning from Sustained Success: How Community-Driven Initiatives to Improve Urban Sanitation Can Meet the Challenges. World Development.

Movik, S. and Mehta, L. 2010. The Dynamics and Sustainability of Community-led Total Sanitation (CLTS): Mapping Challenges and Pathways, STEPS Working Paper 37, Brighton: STEPS Centre.

Muspratt, A.M., Nakato, T., Niwagaba, C., Dione, H., Kang, J., Stupin, L., Regulinski, J., Mbéguéré, M. and Strande, L. 2014. Fuel potential of faecal sludge: calorific value results from Uganda, Ghana and Senegal. Journal of Water, Sanitation and Hygiene for Development 4(2): 223–230. Vancouver.

Naylor, R.L., Hardy, R.W., Bureau, D.P., Chiu, A., Elliott, M., Farrell, A.P., Forster, I., Gatlin, D.M., Goldburg, R.J., Hua, K. and Nichols, P.D. 2009. Feeding aquaculture in an era of finite resources. Proceedings in the National Academy of Sciences. 106: 15103–15110.

Nguyen, H.D. 2010. Decomposition of organic wastes and fecal sludge by black soldier fly larvae. Thailand: Asian Institute of Technology. Pivot Works 2016. http://pivotworks.co/pivot-works.

Quattri, M. and Smets, S. 2014. Lack of community-level improved sanitation causes stunting in rural Lao PDR and Vietnam. WEDC Paper.

Routray, P., Schmidt, W.P., Boisson, S., Clasen, T. and Jenkins, M.W. 2015. Sociocultural and behavioural factors constraining latrine adoption in rural coastal Odisha: an exploratory qualitative study. BMC Public Health 15(1): 880.

Smet, J. and Sugden, S. 2006. Ecological Sanitation, WELL Resource Centre. Loughborough University.

Sanergy. 2016. < http://saner.gy>.

Sanivation. 2016. < http://www.sanivation.com>.

Satterthwaite, David, Diana Mitlin and Sheridan Bartlett. 2015. Is it possible to reach low-income urban dwellers with good-quality sanitation? *Environment and Urbanization* 27.1: 3–18.

SEI. 2015. Managing environmental systems: the water-energy-food nexus. Stockholm Environment Institute: Stockholm. Sweden.

Slamson Ghana Ltd. 2016. <http://www.slamsonghana.com>.

Smet, J. and Sugden, S. 2006. Ecological Sanitation, WELL Factsheet. WELL. Loughborough University, UK.

Strande, Linda. 2014. The global situation in Faecal sludge management: Systems approach for implementation and operation. IWA Publishing.

Strande, Linda, Mariska Ronteltap and Damir Brdjanovic (eds.). 2014. Faecal Sludge Management: Systems Approach for Implementation and Operation. IWA Publishing.

Sugden, S. 2016. The Microbiological Contamination of Water Supplies. WELL Resource Centre.

SuSanA. 2016. < www.susana.org>.

Tilley, E., Ulrich, L., Lüthi, C., Reymond, P. and Zurbrügg, C. 2014. Compendium of Sanitation Systems and Technologies. Dübendorf: Switzerland.

Umande Trust. 2016. < http://umande.org>.

UNEP/UN-HABITAT. 2010. State of water and sanitation in the world cities scales.

UNICEF/WHO. 2014. Progress on Drinking Water and Sanitation 2014. Joint Monitoring Programme. UNICEF/WHO.

UNICEF/WHO. 2017. Progress on Drinking Water and Sanitation: 2017 Update and MDG Assessment. Joint Monitoring Programme. UNICEF/WHO.

United Republic of Tanzania. 2012. Sanitation & Water for All. Briefing: economic impact of water and sanitation. Dar es Salaam: URT.

Van Huis, A., Van Itterbeeck, J., Klunder, H., Mertens, E., Halloran, A., Muir, G. and Vantomme, P. 2013. Edible insects: future prospects for food and feed security (No. 171). Food and Agriculture Organization of the United Nations.

Water and Sanitation Program. 2010. Social Factors Impacting Use of EcoSan in Rural Indonesia, Learning Note.

Water and Sanitation Program. 2014. Investing in the Next Generation. Research Brief.

Water and Sanitation Program. 2016. Multi-sectoral Approaches to Improving Nutrition, Technical Paper.

Waterkeyn, J. and Cairncross, S. 2005. Creating demand for sanitation and hygiene through Community Health Clubs: A cost-effective intervention in two districts in Zimbabwe. *Social Science & Medicine* 61.9: 1958–1970.

WBCSD. 2005. Guidelines for the selection and use of fuel and raw materials in the cement manufacturing process. World Business Council for Sustainable Development.

WRC. 2015. The Status of Faecal Sludge Management in Eight Southern and East African Countries. Report No. KV 340/15. Water Resource Commission: South Africa.

Section 3
Water as a Human Right

Historical Development of Human Rights, Human Rights and Environment and Human Rights to Water

Velma I. Grover

"The love of liberty is the love of others,
The love of power is the love of ourselves." William Hazlitt

"Man is born free
And everywhere he is in chains." Jean-Jacques Rousseau

"Everyone has the right to a standard of living adequate for the health and well-being of himself and of his family, including food, clothing, housing and medical care and necessary social services, and the right to security in the event of circumstances beyond his control." Universal Declaration of Human Rights, United Nations, December 10, 1948

INTRODUCTION

When one looks at the discussion surrounding human rights to water and explores the development of linkages between human rights and environment, one wonders how did human rights come into being. The society has struggled with slavery and colonial rules in the past, and that started the journey of concepts of human rights in the contemporary world. Similarly, exploitation of the environment in the name of development and lack of resources for poor people started a discussion related to human rights to environment and development. The journey of human rights to water can be attributed to the beginning of privatization of water, especially in the developing countries. This chapter begins with a focus on the historical development

981 Main St West, Hamilton, ON, L8S1A8, Canada.

of human rights, followed by how linkages between human rights and the environment were established. The last section of the chapter focuses on water as human rights.

Human Rights and Civil Rights/Civil Liberties

One comes across different terms like Human Rights, Civil Rights and/or Civil Liberties and one wonders whether there is any difference between these terms or these terms are just synonyms. Human Rights are the rights of an individual by just being virtue of being a human and are of recent origin/development, after World War II. Civil rights on the other hand are legally granted for being a citizen, imparted or granted by the constitution of the country. Human Rights are generally considered to be fundamental rights and include the right to education, free expression and free trial. Many of these rights are necessary for human existence. The United Nations General Assembly adopted the Universal Declaration of Human Rights in 1948 cementing their foundation in international law and policy and is widely accepted ever since. Every individual is entitled to certain basic rights, which are either inherent or obtained through the constitution. Human rights and civil rights are the two basic rights that are often debated upon. Both human rights and civil rights have their own features and characteristics.

Civil rights are related to the constitution of each country, whereas human rights are considered a universal right. While human rights are basic rights inherent with birth, civil rights are the creation of society.[1]

After discussing differences between human rights and civil rights, let us explore differences between "civil rights" and "civil liberties". On one hand, civil rights deal with freedom from discrimination or unequal treatment (based on specific themes such as gender, race, and disabilities or settings such as employment or housing), on the other hand civil liberties deal with basic rights and freedoms guaranteed in the Bill of Rights and the Constitution or as unravelled by courts and lawmakers (and include: freedom of speech, the right to privacy).

Human rights are universally accepted rights regardless of nationality and do not change from one country to another, unlike civil rights which differ from one nation to the other. Civil rights basically depend on the laws of the country. Human rights are universally accepted rights regardless of nationality, religion and ethnicity. On the other hand, civil rights fall within the limits of a country's law, and pertain to the social, cultural, religious and traditional standards, and other aspects.[2]

History of human rights

The earliest reference to human rights in the recorded history of mankind can be found way back in 539 B.C. when the armies of Cyrus the Great conquered the city

---

[1] Source: http://www.hg.org/article.asp?id=31546, accessed and http://www.differencebetween.net/miscellaneous/politics/difference-between-human-and-civil-rights/.

[2] (Sources: Civil rights vs. civil liberties: http://civilrights.findlaw.com/civil-rights-overview/civil-rights-vs-civil-liberties.html#sthash.bHUx6MRJ.dpuf;   http://civilrights.findlaw.com/civil-rights-overview/civil-liberties.html and http://www.differencebetween.net/miscellaneous/politics/difference-between-human-and-civil-rights/).

of Babylon and freed the slaves of Babylon by granting them the right to choose their own religion. These decrees together with other similar decrees were recorded on a baked-clay cylinder in the Akkadian language with cuneiform script as shown below:[3]

The Cyrus Cylinder (539 B.C.)

Even today these are known as the Cyrus Cylinder and this ancient record now is recognized as the first charter of human rights. These decrees have also been translated into all six official languages of the United Nations and its provisions parallel the first four Articles of the Universal Declaration of Human Rights.[4] The concept of human rights spread quickly from Babylon to India, Greece and finally Rome. However, further spread was hampered in Rome, because of the prevalence of "natural law" in Rome. One of the reasons could be that natural law in antiquity had justification in a religious reading of the world and because of that natural law suffered unjustified exclusion.[5]

Cylinder—Cyrus' Liberation of Babylon

Clay cylinder used to produce tablets containing the law of Cyrus II (the Great).

The Magna Carta (1215)

Another turning point in the human rights movement was: "Magna Carta", or also known as the "Great Charter" signed by King John of England and his rebellious Barons/subjects (who were lined up against the King at Runnymede in 1215). This probably had the most important impact leading to the rule of constitutional law in the English-speaking world of today. Since the king always violated number of ancient customs and laws of England, his Barons forced him to sign on Magna Carta. In a way parts of Magna Carta were crucial because they subjected the king to the law of the land for the first time in the history of Britain. However, it did not last too long.[6]

Widely viewed as one of the most important legal documents in the development of modern democracy, the Magna Carta was a crucial turning point in the struggle to establish freedoms.

Today it is considered by many researchers to be the blue print for constitutions and bills of rights all over the world. The Magna Carta guarantees certain basic levels

---

[3]   (Source: http://heritageinstitute.com/zoroastrianism/achaemenian/cyrus.htm).

[4]   (Source: http://heritageinstitute.com/zoroastrianism/achaemenian/cyrus.htm and http://www.human rights.com/what-are-human-rights/brief-history/cyrus-cylinder.html).

[5]   (Source: http://www.hprweb.com/2012/06/human-rights-and-natural-law/).

[6]   (Source: http://www.bbc.co.uk/history/british/middle_ages/magna_01.shtml).

of treatment for the individuals and limits the government's ability to act without authority of abuse of its power.

The Magna Carta is the base on which all further institutions were developed and the liberties to free men guaranteed. The Rights were further extended in 1689. John Locke's writings on the nature of government in the late 1600s were inspired by the Magna Carta, who felt that such rights not only belonged to the English, but also to all property-owning adult males.

Perusal of various English parliamentary documents are mainly limited to freeborn Englishmen. The Enlightenment helped broaden the assertions that can also be seen in the American offshoots of the English parliamentary tradition of rights. Declaration of Independence of 1776 claimed that "inalienable" rights were the foundation of all government—thus justified American resistance to English rule.[7]

Documents asserting individual rights, such as the Magna Carta (1215), the Petition of Right (1628), the US Constitution (1787), the French Declaration of the Rights of Man and of the Citizen (1789), and the US Bill of Rights (1791) are the written precursors to many of today's human rights documents.

Petition of Right (1628)

The next milestone in the development of human rights is a document known as the Petition of Rights prepared in 1628 by the English Parliament on civil liberties and sent it to King Charles I.

In response to the King's unpopular policy of exacting forced loans on subjects by quartering troops in their residences as an economy measure, the Petition of Rights was initiated by Sir Edward Coke. It was based on the earlier statutes and charters by asserting four principles: (1) No taxes may be levied without consent of Parliament, (2) No subject may be imprisoned without cause shown (reaffirmation of the right of habeas corpus), (3) No soldiers may be quartered upon the citizenry, and (4) Martial law may not be used in time of peace.[8]

United States Declaration of Independence (1776)

In 1776, Thomas Jefferson penned the American Declaration of Independence.

Declaration of Independence on July 4, 1776, approved by the United States Congress on July 2 was another milestone along the road of human rights. Declaration was a formal account of why the Congress had voted on July 2 to proclaim independence from Great Britain, more than a year after the outbreak of the American Revolutionary War, and as an assertion that the 13 American Colonies were no longer was a part of the British Empire. https://www.encyclopedia.com/history/united-states-and-canada/us-history/declaration-independence.

Impacts of the Declaration of Independence

Philosophically, the Declaration stressed two refrains: individual rights and the right of revolution. The Declaration of Independence gave birth to what is known today as the United States of America. Besides, these ideas spread internationally to other counties as well, inspiring the French Revolution in particular.

---

[7] (Source: https://chnm.gmu.edu/revolution/chap3a.html).

[8] (Source: http://www.bbc.co.uk/history/british/middle_ages/magna_01.shtmlilliers).

The constitution of the United States of America (1787) and Bill of Rights (1791)

One of the well-known examples of a Bill of Rights is the one enshrined in the Constitution of the United States. This automatically grants citizens and residents certain civil liberties—including the right to speak or write freely, to assemble when they want, to practice the religion of their choice and to "bear arms". These rights are frequently relied upon in the American courts.

Basic freedoms of the citizens of the Unites States of America—was the next step in the journey along the Human Rights development—are guaranteed and protected by The Bill of Rights of the US Constitution. The Constitution of the United States of America.

Written during the summer of 1787 in Philadelphia, the Constitution of the United States of America—considered to be the foremost piece of legislation—the fundamental law—limiting the powers of the federal government of the United States protecting the rights of all citizens, residents and visitors to the United States of America. It also outlined a legal framework for legislative system.

Bill of Rights

The Bill of Rights, "protects freedom of speech, freedom of religion, the right to keep and bear arms, the freedom of assembly and the freedom to petition. It also prohibits unreasonable search and seizure, cruel and unusual punishment and compelled self-incrimination." The Bill of Rights prohibits Congress from "making any law respecting establishment of religion and prohibits the federal government from depriving any person of life, liberty or property without due process of law."[9]

Impact of US declaration of Independence on other nations

The US Declaration of Independence left lasting effects upon human rights movement of other foreign nations, such as the French Declaration of the Rights of Man and Citizen, and the Declaration of Independence for the Democratic Republic of Vietnam. The French Declaration was one of the fundamental documents of the French Revolution and defines a set of individual and collective rights of all of the estates as one. The First article states, "(M)en are born and remain free and equal in rights. Social distinction may be founded only upon the general good."? The principles in the French Declaration are still set forth today.[10]

Declaration of the rights of Man and of the citizen (1789)

The Declaration of Rights of Man Citizen in 1789 was another important step along the journey of human rights. The importance of the Declaration of Rights of Man is evident from its comparison the Universal Human Rights passed by the United

---

[9] http://www.humanrights.com/what-are-human-rights/brief-history/declaration-of-independence.htm.
[10] Source: "Effects of the Declaration of Independence." Surfnetkids. *Feldman Publishing*. 28 May. 2008. Web2016.<http://www.surfnetkids.com/independenceday/268/effects-of-the-declaration-of-independence/>.

Nations after World War II in 1948. Both the Documents have similarity—the UN Documents refer to "human beings" instead of "Man" in the 1789 Document.

The Declaration proclaims and guarantees rights of "liberty, property, security, and resistance to oppression" to all citizens. It claims that the need for law stems from the fact that "...the exercise of the natural rights of each man has only those borders which assure other members of the society the enjoyment of these same rights." Thus, the Declaration considers law as an "expression of the general will", intended to promote this equality of rights and to forbid "only actions harmful to the society."

The 18th century movement intended to replace the institutions like hereditary monarchy with institutions based on the principles of the Enlightenment with the aim of applying the methods learned from the scientific revolution to the problems of society. Advocates of such society devoted themselves to "reason" and "liberty"—liberty meant freedom of religion, freedom of press and freedom from unreasonable government interference. Enlightenment writers such as Voltaire, Montesquieu, and Rousseau, became very popular all over western world including the British North American colonies prompting the American revolutionaries to put some ideas in into practice in the Declaration of Independence and the new constitution of the United States of America.

The Declaration of the Rights of Man and citizen of 1789 brought together two streams of thoughts—one stemming from the Anglo-American tradition of legal and constitutional guarantees of individual liberties, and the other from the French revolutionaries who wrote a Declaration of Rights and felt that would serve as a model for the world. Reason rather than the tradition would be the justification.[11]

Life, Liberty, and the pursuit of happiness

The Declaration of Independence gave birth to many other freedoms in the United States of America which may never have even been intended. One of the more immediate effects felt by the Declaration of Independence was the emancipation of black slaves. Abraham Lincoln certainly took literally the statement from the Declaration, "that they are endowed by their Creator with certain unalienable Rights—that among these are Life, Liberty, and the pursuit of Happiness." This in nutshell is the American dream.

The Declaration of Independence also paved the way and created equality among all men and women. Today we can see the effects of the first sentence written in the Preamble: "We hold these truths to be self-evident, that all men are created equal."? Throughout history we have seen so many different changes, from freedom of slaves, to equality among men and women. Today more so than in the past, women have been given every opportunity that men were given and now are truly equal among men.

---

[11]   https://chnm.gmu.edu/revolution/chap3a.html.

Impact upon Bill of Rights

The Declaration of Independence had a profound effect upon the Bill of Rights and the Constitution. The Declaration, it seems, may have ignited the fire under which the Bill of Rights and the Constitution were written. The Declaration is in large part a summary of what the Bill of Rights stands for. The Bill of Rights in the United States is the name by which the first ten amendments to the United States Constitution are known. While the Declaration offered independence from Britain and made general statements, the Bill of Rights offers conclusive and specific rights and laws, from freedom of speech, press and religion, to the right to keep and bear arms; the freedom of assembly; the freedom to petition; prohibits unreasonable search and seizure; cruel and unusual punishment; and compelled self-incrimination. The first ten amendments are truly and expansion on what the first 56 signers of the Declaration had written.[12]

The United Nations (1945) and development of human right laws

World War II raged throughout Europe and Asia, from 1939 to 1945, where cities lay smoldering and in ruins leaving millions of people dead and millions more homeless and starving. Fifty Nations met in San Francisco in 1945 to form United Nations protect and promote peace among nations and thus prevent future wars. The ethics of the organization were detailed in the preamble to its proposed charter: "We the peoples of the United Nations are determined to save succeeding generations from the scourge of war, which twice in our lifetime has brought untold sorrow to mankind."

The Universal Declaration of Human Rights (1948)

The Universal Declaration of Human Rights has inspired other human rights laws and treaties throughout the world. By 1948, the United Nations' new Human Rights Commission had captured the world's attention. Under the dynamic chairmanship of Eleanor Roosevelt—President Franklin Roosevelt's widow, a human rights champion in her own right and the United States delegate to the UN—the Commission set out to draft the document that became the Universal Declaration of Human Rights. Roosevelt, credited with its inspiration, referred to the Declaration as the international Magna Carta for all mankind. It was adopted by the United Nations on December 10, 1948.

In its preamble and in Article 1, the Declaration unequivocally proclaims the inherent rights of all human beings: "Disregard and contempt for human rights have resulted in barbarous acts which have outraged the conscience of mankind, and the advent of a world in which human beings shall enjoy freedom of speech and belief and freedom from fear and want has been proclaimed as the highest aspiration of the common people. ...All human beings are born free and equal in dignity and rights."

---

[12] Source: "Effects of the Declaration of Independence." Surfnetkids. *Feldman Publishing*. 28 May, 2008. Web. 15 May, 2016. <http://www.surfnetkids.com/independenceday/268/effects-of-the-declaration-of-independence/>.

The Member States of the United Nations pledged to work together to promote the 30 articles of human rights that, for the first time in history, had been assembled and codified into a single document. In consequence, many of these rights, in various forms, are today part of the constitutional laws of democratic nations.[13]

We have tracked the history of development of human rights. The next section looks at the development of environmental laws and environmental rights as human rights followed by development of water as a human right.

Development of international environmental laws

In the 19th century there was some interest in conserving environment mainly wildlife, fisheries, birds and seals. With industrialization, the need for natural resources grew and so did conflicts over natural resources, which started an era of use of international arbitration to settle environmental disputes. The first two cases that stand out are: Fur Seal Arbitration (between the US and Great Britain—where Great Britain was over-exploiting fur seals beyond its territorial jurisdiction) and the Trail Smelter Arbitration Case (between the US and Canada, where nitrous fumes from Canada were impacting the American State of Washington). In the first case, it was recognized "the right of a State to exploit its resources, this case affirms the obligation to act in good faith and a related prohibition on the exercise of one's rights solely to cause injury to another (p. 12)."[14] In the second case, the tribunal stated: 'no State has the right to use or to permit the use of its territory in such a manner as to cause injury ... to the territory of another or the properties or persons therein (p. 12). ...'[15] This case of 1941 led to the establishment of the "Polluter Pays Principle".

After World War II and the creation of the UN, specialized UN agencies such as Food and Agricultural Organization (FAO) was created followed by others such as United Nations Economic, Social and Cultural Organization (UNESCO) and International Union for the Conservation of Nature (IUCN). Due to growing concern for environment and clear recognition between conservation and development, the very first international conference was convened in 1949.[16] The next major milestone in the journey of environmental laws is the 1949 decision of International Court of Justice: Corfu Channel Case, which reaffirms State's obligation not to "allow knowingly its territory to be used for acts contrary to the rights of other States."[17] This case (later known as Lac Lanoux Arbitration) applied the principle of riparian rights in the context of shared rivers. The Biosphere Conference organized by UNESCO

---

[13] Source: http://www.humanrights.com/what-are-human-rights/brief-history/the-united-nations.html).

[14] Asia Pacific Forum, Human Rights and the Environment, Background paper, APF 12, the 12th Annual Meeting of the Asia Pacific Forum of National Human Rights Institutions, Sydney, Australia, 24–27 Sep. 2007.

[15] Asia Pacific Forum, Human Rights and the Environment, Background paper, APF 12, the 12th Annual Meeting of the Asia Pacific Forum of National Human Rights Institutions, Sydney, Australia, 24–27 Sep. 2007.

[16] UN Conference on the Conservation and Utilization of Resources convened by UN Economic and Social Council.

[17] p. 13, Asia Pacific Forum, Human Rights and the Environment, Background paper, APF 12, the 12th Annual Meeting of the Asia Pacific Forum of National Human Rights Institutions, Sydney, Australia, 24–27 Sep. 2007.

in 1968 brought scientists around the world together to discuss human impact on the environment, and the final report of this conference led to the most recognized milestone in the contemporary international environmental laws: the Stockholm Conference.[18]

The Stockholm Conference in 1972 (UN Conference on the Human Environment) brought attention of the international community to linkages between the environment and development and led to creation of United Nations Environmental Program (UNEP), Organization for Economic Cooperation and Development (OECD) and the Declaration of Principles for the Preservation and Enhancement of the Human Environment. The Declaration has 26 principles which emphasize the need of preservation of environment but also balancing both natural and man-made environment for enjoyment of basic human rights. The very first declaration states:

"Man has the fundamental right to freedom, equality and adequate conditions of life, in an environment of a quality that permits a life of dignity and well-being, and he bears a solemn responsibility to protect and improve the environment for present and future generations."[19] For the purpose of this discussion, Principle 1 on human right to the environment is of most interest:

"All human beings have the fundamental right to an environmental adequate for their health and well-being"[20]

This inspired some countries to enshrine this in their constitutions. For example, India amended its constitution and added Article 48-A[21] and Article 51A(g).[22] Article 48 (A) puts the responsibility on the State to protect environment and states: "The State shall endeavour to protect and improve the environment and to safeguard the forests and wildlife of the country" while Article 51A(g) also calls on every citizen to protect the environment: "It shall be the duty of every citizen of India to protect and improve the natural environment including forests, lakes, rivers and wildlife and to have compassion for living creatures."[23]

On the international level, efforts were made to bring more linkages between human rights and the environment. For example, linkages between Indigenous Peoples and their environment was recognized by the International Labour Organization's Convention No. 169. The Article 7(3) of this Convention clearly

---

[18] Asia Pacific Forum, Human Rights and the Environment, Background paper, APF 12, the 12th Annual Meeting of the Asia Pacific Forum of National Human Rights Institutions, Sydney, Australia, 24–27 Sep. 2007.

[19] p. 13/14 Asia Pacific Forum, Human Rights and the Environment, Background paper, APF 12, the 12th Annual Meeting of the Asia Pacific Forum of National Human Rights Institutions, Sydney, Australia, 24–27 Sep. 2007.

[20] p. 16 Asia Pacific Forum, Human Rights and the Environment, Background paper, APF 12, the 12th Annual Meeting of the Asia Pacific Forum of National Human Rights Institutions, Sydney, Australia, 24–27 Sep. 2007.

[21] Falls under Directive Principles of State Policy.

[22] Comes under Fundamental Duties.

[23] http://www.environmentallawsofindia.com/the-constitution-of-india.html and Dr. Ojha, K.B. 2013. Human right and environment pollution in india and judiciary contribution. *International Journal of Humanities and Social Science Invention* ISSN (Online): 2319–7722, ISSN (Print): 2319–7714. www.ijhssi.org Volume 2 Issue 11 I November. I pp. 42–47. Accessed from http://www.ijhssi.org/papers/v2(11)/Version-1/H021101042047.pdf.

states: "Governments shall take measures, in cooperation with the peoples concerned, to protect and preserve the environment of the territories they inhabit" (International Labour Organisation Convention No. 169 concerning Indigenous and Tribal Peoples in Independent Countries, 1989).[24] The Hague Declaration on the Environment in 1989 explicitly recognizes the connection between human rights and the environment in its first paragraph: "The right to live is the right from which all other rights stem. Guaranteeing this right is the paramount duty of those in charge of all States throughout the world".[25]

Although the Stockholm Conference clearly links human rights to the environment, the Rio Conference (1992) fails to develop on it further—in fact it just uses a broad brush still connecting environment and development through various principles. However, the Rio Conference did reinforce the Precautionary Principle and the Polluter Pay Principle and made some progress towards right to access to information.[26]

To conclude this section, it can be said that there have been a lot of effort to link the environment to human rights but there is no clear articulation of the right to environment in any international human rights treaties such as: Universal Declaration of Human Rights or International Bill of Rights. At the local or regional level, The African Banjul Charter on Human and People's Right 1981 (African Charter) and the American Convention on Human Rights in the Area of Economic, Social and Cultural Rights (1988) do contain a specific environmental right.[27] The African Charter in Article 24, states:[28] "All peoples shall have the right to a general satisfactory environment favourable to their development". Although, in the African system, there is a limited jurisprudence on the right to environment yet in the case of "The Social and Economic Rights Action Centre and the Centre for Economic and Social Rights *v.* Nigeria, that Article 24 obliges States to: take reasonable and other measures to prevent pollution and ecological degradation, to promote conservation, and to secure an ecologically sustainable development and use of natural resource".[29]

---

[24] http://pro169.org/res/materials/en/general_resources/Manual%20on%20ILO%20Convention%20 No.%20169.pdf and Asia Pacific Forum, Human Rights and the Environment, Background paper, APF 12, the 12th Annual Meeting of the Asia Pacific Forum of National Human Rights Institutions, Sydney, Australia, 24–27 Sep. 2007.

[25] 1989 Hague Declaration on the Environment, 11 March 1989, 28 I.L.M. 1309 (1989) and Asia Pacific Forum, Human Rights and the Environment, Background paper, APF 12, the 12th Annual Meeting of the Asia Pacific Forum of National Human Rights Institutions, Sydney, Australia, 24–27 Sep. 2007.

[26] Asia Pacific Forum, Human Rights and the Environment, Background paper, APF 12, the 12th Annual Meeting of the Asia Pacific Forum of National Human Rights Institutions, Sydney, Australia, 24–27 Sep. 2007.

[27] Asia Pacific Forum, Human Rights and the Environment, Background paper, APF 12, the 12th Annual Meeting of the Asia Pacific Forum of National Human Rights Institutions, Sydney, Australia, 24–27 Sep. 2007.

[28] African [Banjul] Charter on Human and Peoples' Rights, adopted on 27 June 1981, OAU Doc. CAB/ LEG/67/3 rev 5, 21 I.L.M. 58 (1982), entered into force Oct. 21, 1986.

[29] The Social and Economic Rights Action Centre and the Centre for Economic and Social Rights *v.* Nigeria, African Commission on Human and Peoples' Rights, Comm. No. 155/96 (2001) at para accessed from http://www1.umn.edu/humanrts/africa/comcases/155-96.html AND P. 27 Asia Pacific Forum, Human Rights and the Environment, Background paper, APF 12, the 12th Annual Meeting of the Asia Pacific Forum of National Human Rights Institutions, Sydney, Australia, 24–27 Sep. 2007.

The American Convention on Human Rights in the Area of Economic, Social and Cultural Rights in Article 11 states: "Everyone shall have the right to live in a healthy environment and to have access to basic public services.

The States Parties shall promote the protection, preservation, and improvement of the environment."[30]

The major milestone in linking the environment and development was clearly seen by the Stockholm Conference but the most comprehensive effort to mainstream human rights into development approach began in 1997 under the direction of Kofi Annan.[31] The international community has accepted the relationship between the two as clearly demonstrated by statements of Judge Weeramantry (International Court of Justice) and Klaus Toepfer (Executive Director of UNEP). Judge Weeramantry said:

> "the protection of the environment is ... a vital part of contemporary human rights doctrine, for it is a sine qua non for numerous human rights such as the right to health and the right to life itself ... [D]amage to the environment can impair and undermine all the human rights spoken of in the Universal Declaration and other human rights instruments."[32]

While Klaus Toepfer stated:[33]

"The fundamental right to life is threatened by soil degradation and deforestation and by exposures to toxic chemicals, hazardous wastes and contaminated drinking water. Environmental conditions clearly help to determine the extent to which people enjoy their basic rights to life, health, adequate food and housing, and traditional livelihood and culture."

Water: a human right or a commodity[34]

Development of human right to the environment was the first milestone towards water as a human right. The fundamental question that arises here is, whether water is a commodity, a service which you can have only, if you can afford it, or is it a human right which everyone should have (at least a certain minimum quantity) to live with dignity. The argument that water is a commodity has led to privatization of some or part of water delivery system controlled by multinational corporations

---

[30] Additional Protocol to the American Convention on Human Rights in the Area of Economic, Social and Cultural Rights (the Protocol of San Salvador), Adopted at San Salvador, 17 November 1988, OAS Treaty Series 69. AND P. 27 Asia Pacific Forum, Human Rights and the Environment, Background paper, APF 12, the 12th Annual Meeting of the Asia Pacific Forum of National Human Rights Institutions, Sydney, Australia, 24–27 Sep. 2007.

[31] Human Rights and Development: towards mutual reinforcement. Edited by Philip Alston and Mary Robinson: Center for Human Rights and Global Justice. New York University, School of Law, 2005.

[32] P. 85 Asia Pacific Forum, Human Rights and the Environment, Background paper, APF 12, the 12th Annual Meeting of the Asia Pacific Forum of National Human Rights Institutions, Sydney, Australia, 24–27 Sep. 2007.

[33] P. 85 Asia Pacific Forum, Human Rights and the Environment, Background paper, APF 12, the 12th Annual Meeting of the Asia Pacific Forum of National Human Rights Institutions, Sydney, Australia, 24–27 Sep. 2007.

[34] Velma I. Grover. 2010. Water: A Human Right or a Commodity? Seronica *The Journal of Socio Environmental Research Organization*. Vol. 1 No. 3.

(MNCs). In turn, there has been an increased concern that the poor are left out because the MNC's main responsibility is to increase profit as they are accountable to shareholders. As a result, many people may not be able to afford water, because they are too poor to pay for it. For example, a family in Bolivia living just behind a water plant could not afford to get a water connection, because they would have to pay nine months salary just to get a water connection, and the monthly charges are on top of it—highlighting the issue of access inequality in case of privatization of water. This raises the question of access to safe drinking water as a human right. According to the World Health Organization (WHO) each citizen should be assured 20 liters of water by the governments. While Article 25 of the United Nations Declaration of Human Rights does not mention water explicitly, yet it does mention the right to food, which includes water. Water is essential for humans to live. The access to water should, therefore, be a basic/fundamental right with universal access.

Water is one of the vital resources for all life on earth—especially for the human life. The availability and quality of water always played an important role in determining where people live and the quality of life would be (which is why most of the earlier settlements were along the coast or the rivers). Safe drinking water and access to water is important for overall human well-being. According to the definition of WHO safe drinking water implies that the water meets accepted drinking water quality standards and poses no significant threat to health.

Demand for water comes from various sectors such as agriculture, industry, hydro-power plants, recreation, domestic needs, etc. Availability of water in certain areas is also being complicated by climate change, which is not only resulting in extreme weathers but also is changing the hydrological cycle and precipitation patterns adding to water stress in some areas. Today, about a billion people lack access to safe drinking water and millions die a year due to water borne diseases. This is all preventable by simple access to safe drinking water. A minimum quantity of 20 litres of water per person per day as stipulated by UN agencies such as UNDP and WHO is considered a natural right of people and part of overall human right of each citizen of a state.

The UN Convention on the Rights of the Child in 1989 took a step further in recognizing the impact of the environment on human rights. Although, the convention does not include a specific human right to the environment but Article 24(2)(c) of the convention mentions clean drinking water and also calls for measures to combat disease and malnutrition:

> "through the provision of adequate nutritious foods and clean drinking water, taking into consideration the dangers and risks of environmental pollution."[35]

The Article goes on to require the provision of information and education to all segments of society on hygiene and environmental sanitation (Art. 24(2)(e)).

---

[35] P. 17 Asia Pacific Forum, Human Rights and the Environment, Background paper, APF 12, the 12th Annual Meeting of the Asia Pacific Forum of National Human Rights Institutions, Sydney, Australia, 24–27 Sep. 2007.

A key document regarding human rights to water is General Comment No. 15 (GC 15) of the Economic and Social Council (ECOSOC) proclaiming the human right to water. This agrees with Articles 11 and 12 of International Covenant on Economic, Social and Cultural Rights (ICESCR), which affirms that the human right to water entitles everyone to "sufficient, safe, acceptable, physically accessible and water for personal and domestic uses". A fundamental weakness of right to water as recognized in General Comment No. 15 is that it is an independent right founded on non-binding instrument. Another criticism of GC 15 is that while it claims the existence of the independent right to water, it does only by reference to other rights including the right to health, the right to an adequate standard of living and the right to food.

Although GC 15 is not binding, the document can be seen as a departure from international law by explicitly recognizing a right to water either as a derivative right or an independent right, but at least it represents a standing source of legal interpretation. The first obvious right associated with the provision of clean water and sanitation facilities is the right to the highest attainable standard of health. On a similar level, the right to an adequate standard of living dependent on the supply of clean and safe water and proper sanitation services.

However, the development of GATS (General Agreement in Trade in Services) may interfere with basic human rights, especially the right to access of safe drinking water. In a situation of global water scarcity, many consider that the privatization of essential services following the liberalization will impede the provision of water to everyone at affordable prices. For instance, the human rights doctrine prescribes core obligation that countries are bound to respect and fulfill, without derogation, primary health care, food, access to water, basic shelter and housing, right to life, right to be free from torture. On the other hand, the multilateral trading systems aim to encourage trade by the reduction of trade barriers. The barriers can take the form of custom duties, but also, particularly for trade in services, the form of regulatory measures. The principles of trade law, including non-discrimination, reduction of quantitative restrictions to trade and transparency requirements, limit the state regulation. Essentially when a private company comes in to build and operate a system in a country it invests in infrastructure and wants to recover the cost (in addition to making a profit) and when the commodity involved is water, there is a huge amount of investment required in infrastructure but at times the full recovery costs in poor areas make it difficult for the most needy to get access to the services.

UNGA action

In spite of reference to "human right for access to safe drinking water and sanitation" in various international instruments, Bolivia felt that these rights should be formally recognized by the UN General Assembly especially before the upcoming summit to review progress on The Millennium Development Goals. The proposal put forward by Bolivia was put to vote by the UN General Assembly on July 28th July 2010 declaring—"the right to safe and clean drinking water and sanitation [is] a human right that is essential for the full enjoyment of life and all human rights." One hundred and twenty two countries cast their vote in favour of the resolution with 41 abstentions. No country opposed the resolution.

The resolution is definitely a milestone. While legally non-binding, such a statement by the highest international assembly indicates that the notion of water as a human right is gaining momentum. At the very least, it adds moral (and potentially political) weight to the belief that governments have a responsibility to ensure safe, clean, accessible and affordable drinking water and sanitation, at least for their own citizens. It was not a conclusive or binding resolution, but a step in the right direction.

CONCLUSION

The development of the specific right to water is the answer of the international community to emergence of water related issues. In view of the movement toward the recognition of a right to water, it could be expected that the right to water will be crystallized in future in a binding form. This is more important for the people who currently lack access to safe drinking water and sanitation than people who can afford to have access to safe drinking water and sanitation.

Some privatization programs have produced positive results. But the overall record is not encouraging. The conviction that the private sector offers a solution for attaining the equity and efficiency needed to accelerate progress towards water for all has proven to be misplaced. While the past failures of water concessions do not provide evidence that the private sector has no role to play, they do point to the need for greater caution, regulation and a commitment to equity in public-private partnerships.

Chapter-10

Justiciability and the Case for Constitutionalization

Amani Alfarra,[1,]* *Islam Abdelgadir*[2] and *Velma I. Grover*[3]

INTRODUCTION

This chapter argues that the judicial enforcement of the duty to fulfil the right to water is appropriate at the domestic level, and that the optimal strategy moving forward is the explicit recognition of the right to water in national constitutions— or 'Constitutionalization'. Taking water as indicative of wider Economic, Social and Cultural Rights (ESCR); Section 1 outlines the contemporary discourse around the right to water and highlights the need to focus on the duty to fulfil the right to water. Critics of this position view that the right to water should best be understood as a directive principle for the long-term guidance of executives and legislatures. This chapter normatively argues that directive principles, whilst necessary, are insufficient for states to fully discharge their obligations. Section 2 further elaborates on these arguments and addresses non-justiciability concerns around the right to water and the complexities posed by limited resources. Looking at the judicial review at the national level, Section 3, makes a positive case for the explicit recognition of the right to water in national constitutions with reference to relevant case law in South Africa and Uruguay.

Basic Framework for the Right to Water

Water as an Economic, Social and Cultural Rights (ESCR)

Water was first given authoritative definition as a human right in 2002, when the UN Committee on Economic, Social and Cultural Rights (CmtESCR) issued General

[1] Water Resource Officer, CBL Division --FAO of the U.N., Viale delle Terme di Caracalla, 00153 Rome, Italy.
[2] MA Human Rights, University College London, 7A Herbert Road, NW9 6AJ, London, United Kingdom. Email: islam.abdelgadir@gmail.com
[3] Adjunct Professor, York University.
* Corresponding author: amani.alfarra@fao.org

Comment No. 15. Since the International Covenant on Economic, Social and Cultural Rights (ICESCR) does not explicitly define the rights, the CmtESCR derived the right from Articles 11 and 12—namely, the rights to *"an adequate standard of living"* and *"the right to the highest attainable standard of...health"*. General Comment 15 proclaims that: *"The human right to water entitles everyone to sufficient, safe, acceptable, physically accessible and affordable water for personal and domestic uses"* (para. 2).

Discourse around the human right to water has developed substantially since 2002, with 23 bodies working on water trying to unify under UN-Water and also integrating human rights into development approach (UN Secretary-General, 2003: para. 9). However, a lack of legal consolidation has inhibited its full realization. Even in 2015 663 million people still lacked access to clean, potable water (JMP, 2015). Objections to legal recognition of the right to water are based on a lack of justiciability and its inception as a 'derivative right' (Bulto, 2011). This chapter will focus primarily on the justiciability objection—an issue symptomatic of wider ESCRs. Non-justiciability refers to the difficulty with which the right can be clearly and comprehensively expressed in International Human Rights Law (IHRL). This lack of clarity impedes the identification of rights-holders and duty-bearers, making it difficult for courts to remedy human rights violations (Tushnet, 2008).

International human rights obligations

Criticisms of the right to water's legal basis are well reflected in the debate between ESCR and Civil and Political Rights (CPR). CPR are widely conceived as 'liberty' rights—the right to privacy, the right against cruel and degrading punishment—these rights impose largely negative duties. ESCR represent 'entitlement' rights (Coy et al., 2008)—imposing largely positive duties. Water's necessity for human life and survival, makes it a primary good, and its relation to basic subsistence speaks directly to Shue's (1996) and Rawl's (2001) concerns for the fulfillment of ESCR. Whilst being interpreted as an ESCR, General Comment 15 identifies both negative and positive duties for states in relation to the right to water. These duties manifest themselves within the *'Respect, Protect and Fulfill'* doctrine (para. 20).

The duty to "Respect and Protect" considers negative rights, such as the right to be free from water contamination or arbitrary disconnection. The "duty to fulfil" requires states to take affirmative measures providing individuals with the opportunity to satisfy those needs which they are unable to secure themselves. This can be achieved through discharging their obligations to facilitate, to promote and to provide (ICESCR, 1996: para. 25). In paragraph 10, the committee highlights this specifically as an *"entitlement...to a system of water supply and management"*. This entitlement is one towards a particular kind of economic and social order (Donnelly, 2003), raising concerns of whether fulfilling this entitlement through legal procedure is entirely appropriate. These concerns are true of wider ESCR. As such, this chapter will focus on judicially enforcing the duty to fulfil the right to water in order to make the more substantive case.

Programmatic framework

A counterargument to the disagreement in this chapter would be that the general comment's recognition of the right to water should exist solely as a framework for guiding long-term programs. It is crucial to emphasize that General Comments are only interpretations of the ICESCR by the CmtESCR, whose principle function is to monitor implementation (Tully, 2005). Due to the wide range in economic development and sociocultural constitution among the States party to the ICESCR, the States are given a wide margin for implementation because of different resources available to them. Indeed, the special rapporteur on the human right to water argued that the human right to water does not entail "*any particular form of service provision*" (De Albuquerque and Roaf, 2012). States can discharge their legal obligations in a progressive manner rather than an immediate manner, using their "*maximum available resources*" (ICESCR, 1966: Article 2). In a critique of General Comment 15, Phillipe Cullet argues that the "*expert nature of the committee precludes it being more than a... source of inspiration*" (2013). For the States party to the Covenant, there would seem to be no obligation to establish formal complaint mechanisms in order to discharge their obligations to fulfil the right to water, as there is no requirement under IHRL.

Critics of judicial review can use this counterargument to label it as inappropriate. The following section will make the case that judicial enforcement of the right to water is in fact appropriate by addressing resource requirements, justiciability concerns and identifying states' immediate obligations.

The Case for Judicial Review

Resource requirements

One of the core difficulties in the context of ESCR is the amount of resources required should the States sincerely aim to discharge their obligations. This difficulty is directly applicable to the right to water, and highlights concerns around resource allocation. Some would argue that maybe fiscal authority should be given to courts but it raises questions of democratic deficit. Moreover, whilst judges specialize on legal matters, they may not be best suited to adjudicate on specific ESCR provision such as levels of investment in water infrastructure. They may also be ill-equipped to decide the extent of public/private provision or how to resolve trade-offs between providing for more than one ESCR (Australia, 2009).

Yet the right to water is not unique in demanding resources and carrying implications for resource allocation. Court judgements often have resource implications. The South African case of *Soobramoney vs. Minister of Health* found that judiciaries can take into account resource concerns. Moreover, in the case of *KMS vs. West Bengal*, the Indian court found that it cannot refuse legal aid on account of "*financial constraints*" (Agrawal Judgement para. 16). Courts regularly balance conflicting CPRs, and it is unclear why their judgement cannot be extended to ESCRs.

One should be cautious before dismissing resource concerns entirely—distinguishing the "*incidental costs*" of protecting CPR from the more "*substantial cost of economic redistribution*" is prudent (Neier, 2006). However, the objective here is to show that judicial review can be contextually appropriate.

Non-justiciability

General Comment 15 is vague in defining a 'clean', 'reliable' or 'sufficient' water supply. Defining the scope of states obligation when fulfilling the right to water involves addressing complicated questions of content and criteria (Dennis and Stewart, 2004). A lack of clarity can lead to an understanding of states obligations as inherently non-justiciable.

This logic creates a de facto hierarchy of justiciable CPR over non-justiciable ESCR—contravening the idea that the two sets of rights as '*indivisible, interdependent and interrelated*' (Vienna Declaration: Article 1(5)). The integral role of water as basic subsistence means that lack of access to clean and safe drinking water can undermine an individual's capacity to realize civil and political rights (Sen, 2009). Similarly, Jean Dreze (2004) highlights that many people are incapable of democratic participation due to disempowerment—a condition which water insecurity exacerbates. Indeed, these views are substantiated by the committee itself in Article 1 of General Comment 15, where water is described as a "*prerequisite*" for realizing other rights. Consequently, the CmtESCR General Comment 9 states explicitly that automatically excluding judicial enforcement of ESCR would be "*arbitrary and incompatible with the principle of indivisibility and interdependence*" (para. 10).

Immediate obligations to fulfil

Non-justiciability proponents argue that the relative open-endedness of the concept of "*progressive realization*" renders obligations devoid of any meaningful content (Salman and McInerney-Lankford, 2004). Yet identifying the immediate obligations present in the duty to fulfil can illustrate clearly defined requirements, and be used as the basis for justiciable claims. As such, failing to discharge these immediate obligations can be understood as violating the said right. With reference to the Limburg principles on the implementation of the ICESCR—this chapter highlights three of those violations: neglecting minimum core thresholds, discrimination and retrogression.

States must ensure a *minimum core* threshold of each right (Young, 2008). This minimum core may vary accordingly based on their socioeconomic history but includes at least the minimum decencies of life consistent with human dignity. Although General Comment 15 provides little indication of how a 'minimum' water supply could be measured, the Committee defers to guidelines developed by the World Health Organisation for national standards (Tully, 2005). If a significant proportion of a population fall under these standards, then the state is "*prima facie, failing to discharge its obligations*" (CESCR, 2003: para. 10).

States cannot discharge their obligations in a way that is discriminatory to any individual or group in society (CESCR, 2009). In *Mazibuko vs. City of Johannesburg*, the court rejected pre-paid water meters on the basis that they were racially discriminating. Indeed, the Limburg principles require states parties to ensure the provision of judicial review should discrimination occur (1986, para. 35).

Through 'progressive realization', states must show that they are taking "*deliberate, concrete and targeted*" steps (CESCR, 1990: para. 2). This may not in itself involve judicial review. Many countries formulate water strategies and compare water indicators to UN development levels. These can be sufficient if it can be justified that a state is moving in the least restrictive manner. Under the CmtESCR's interpretation of state's ESCR obligations (1990, para. 9): deliberate retrogression on fulfilling the right is unacceptable without careful justification, and state non-compliance can form the basis of a legal challenge.

Directive principles are insufficient

The true value of the right to water lies in its effective implementation (Gleick, 1998). CmtESCR in its General Comment 9 (1998) argues that states have a duty to give effect to the ICESCR in the domestic legal order using "*all means at (their) disposal*". Although realizing the right to water through programmatic frameworks can be a powerful tool for discharging obligations to fulfil, establishment of accountability mechanisms through judicial oversight can be crucial for ensuring that the progresses is made beyond rhetoric. The principle of *Effet Utile* supports reading international treaties in a manner designed to best give effect to their provisions (Alston and Goodman, 2012). Practically speaking, judicial oversight can drastically reduce the likelihood of discrimination or state retrogression.

While researching Peri-Urban Delhi, Mehta et al. highlight that legislative methods can be hampered by "*elite biases, political economy and jurisdictional ambiguities*" (2014)—stifling distributive justice which may be better achieved by legal institutions. Political capture and a lack of political will can seriously impede the realization of "human right to water". Additionally, due to the expansion of water provision through private means, non-state actors, the influence of private sector is increasing which can substantially impact water legislation. That being said, the delicate role of water among differing sociocultural contexts may be better navigated by judicial authorities (Swyngedouw et al., 2014).

Nonetheless, caution must be displayed before recommending an "*overreaching legal positivism*" (Dennis and Stewart, 2004) assuming that judicial processes produce more desirable outcomes than legislative ones. Indeed, in *Manqele vs. Durban Transitional Metropolitan Council* (2002), South Africa judged that without legislative guidance defining the content, the right to water is incomplete and unenforceable. Ultimately, fulfilling the right to water requires a wide variety of state action—in which judicial review can be appropriately recommended.

The Case for Constitutionalization

The authors also take the position that judicial enforcement is more effective in national settings as opposed to international settings. It is individual member states who are party to the ICESCR and who are obligated to fulfil the right. This chapter does not advocate ignoring international complaints mechanisms, but chooses instead to focus on how the judicial framework can be optimally developed in relation to the executive and legislative branches of the governments.

In domestic courts, enforcing the right to water has been pursued through a plethora of means. Some states follow the interpretation of the CmtESCR and interpret justiciable CPR to encompass the right to water. For example, the Indian Supreme Court has interpreted "Right to Life" covered under Article 21 of the constitution to include "right to water" as a fundamental human right (Winkler, 2008; Kothari, 2006). Although the Indian Constitution fails to recognize a human right to water, yet the judiciary recognizes it entrenched under Article 21 of the Constitution. In addition to the legal precedence set by Indian courts to recognize water as a fundamental human right, a number of states have passed legislations to realize this right and even the Union government has adopted some policy instruments to realization of the right (especially in rural areas) (Cullet, 2013). This has created a gap where legally right to water exists but in terms of the constitution there is no clear definition or content on what falls under the right to water (Cullet, 2013). This leaves the door open for interpreting the right to water in different ways by different courts and can be misused as well. In some cases, the governments are using the discourse of human rights as a rationale for large scale infrastructure (Radonic, 2015). For example, as discussed by Radonic (2017), in Sanora, "the human right to water became a countermobilization by state authorities to the right to water framework that was advanced by indigenous leadership". Radonic (2017) presents an analysis of the case law in Mexico, which shows that indigenous rights intersect with the right to water, that has been derived from Article 4 (1),[1] Article 21 (1)[2] and Article 5[3] and the Court has offered an evolutionary interpretation of these articles to reflect changes in society (Radonic, 2017). Radonic has come to a similar conclusion in Colombia where "Columbia's constitutional court, inter American Court, indigenous reality and diversity are gradually shaping and influencing a multicultural interpretation of human rights" (Radonic, 2017).

A growing number of States have explicitly recognized the right to water by adopting it in their national constitutions—or adopted 'Constitutionalization'. The authors believe that this method provides the optimal framework for judicially enforcement. Through Constitutionalization, the human right to water is established as an independent right—unattached to existing rights of health or life. When a constitution provides that level of legislative specificity, it is more effective to identify and remedy human rights violations. The States that have followed Constitutionalization have witnessed a proliferation of case law on the subject—

---

[1]   The right to life.
[2]   Right to property.
[3]   The right to human dignity.

further clarifying the remit of their judiciary. Cullet argues that the right to water is more effectively enforced when it is clearly defined in legal instruments and given an *"uncontested basis in national law"* (2013). Amnesty International has remarked that Slovenia's recent constitutional amendment guaranteeing the right to water has *"strengthened the case of anyone challenging their lack of access to water in domestic courts"* (Amnesty International, 2016).

Through Constitutionalization, the judiciary is obliged to hold legislative and executive action to account. Legislation can be deemed as unconstitutional if it fails to fulfil the right to water or commits retrogression, and an emboldened judiciary can better identify those concrete and deliberate steps needed for progressive realization.

Varun Gauri highlights that Constitutionalization can still lend itself to significant deference as *"courts play a variety of roles in different settings"* (2009). South Africa, Ecuador and Kenya inter alia have all adopted Constitutionalization but do not specify the need for direct public provision (Barlow, 2008). Looking at the South African context, Constitutionalization has allowed courts to deliver their own interpretation of the right to water. An interpretation which, unlike that of General Comment 15, is legally binding. With reference to the right to water, the state is required to take *"reasonable and other measures..., to achieve progressive realization"* according to Section 27(2) of its 1996 Constitution. This concept of *'reasonableness'* was used to identify the scope of enforceable duties imposed on the state, overcoming traditional non-justiciability concerns.

In Uruguay, a referendum declaring water a constitutional right has provided tools for social movement activists to challenge national courts. The right to water is particularly poignant in this region due to its connections with environmental justice and similar initiatives have been pursued in Bolivia, Ecuador and Mexico (Davidson-Harden et al., 2011). Conversely, the legal scholar Tamar Meshel (2015) argues that these mechanisms can be undermined by investor-state arbitration which often supersedes IHRL in protecting investment interests over human rights.

The bluntness of enforcement mechanisms highlights one vulnerability of Constitutionalization. Judicial mechanisms are often largely dependent on quality of governance. If a judiciary is corrupt or unwilling to recognize the right to water in a broad sense, then it's practical impact may be limited (Rosenburg, 2008). Furthermore, Alston and Goodman argue that if *'rights-consciousness'* is poor, then individuals may not fully take advantage of existing complaints mechanisms (2012). One could look to the judicial activism of the Indian High Court and its aggressiveness as an alternative. However, the strength of the Constitutionalization position lies in the protection it provides from legislative encroachment. A state which fails to incorporate the right to water in a meaningful way into its constitution may never truly be able to guarantee its realization.

The launch of water rights into mainstream political discourse could help to expand rights consciousness. Since the publication of General Comment 15, the number of states recognizing the right to water has doubled in 2010 (Barlow, 2015) and the UN officially recognized the right to water and sanitation through resolution 64/292. It could even be feasibly argued that the evolution of constitutionalization in a growing number of states has proved self-reinforcing, perhaps granting support for its elevation into customary international law.

Conclusion

Since the recognition of the right to water in General Comment 15, the expanding literature on the human right to water has been accompanied with diverging legal institutions. Attempts at judicial enforcement have been met with criticisms of the right's non-justiciability, in a manner reminiscent of the criticisms besetting wider ESCR. This chapter is as attempt to state that the right to water as specified under General Comment 15 entails immediate obligations to fulfil those that are directly justiciable and that it can be appropriate to organize judicial review alongside action by legislative and executive means. This chapter has also made the case that explicitly recognizing this right in national constitutions is the best approach to ensuring that the duty to fulfil this right is effectively realized.

The right to water has many characteristics in common with other ESCR, making much of the analysis here reproducible when considering other rights. One could even argue that its inception initially as an interpretation of CmtESCR's has exacerbated conventional concerns of ambiguity. However, water's importance to our basic subsistence separates it from other rights in the ICESCR such as "*the right to cultural life*" outlined in article 15. Further research could clarify states' obligations to fulfil the right internationally or better examine how fulfilling the right to water can conflict with environmental justice.

Bibliography

International Treaties

1986 Limburg Principles on the Implementation of the International Covenant on Economic, Social and Cultural Rights (1987).

CESCR. 1990. General Comment No. 3: The Nature of States Parties' Obligations (Art. 2, Para. 1, of the Covenant). https://www.refworld.org/docid/4538838e10.html.

CESCR. 2003. General Comment No. 15: The Right to Water (Arts. 11 and 12 of the Covenant) (https://www.refworld.org/docid/4538838d11.html).

CESCR. 2009. General comment no. 21, Right of everyone to take part in cultural life (art. 15, para. 1a of the Covenant on Economic, Social and Cultural Rights). https://www.refworld.org/docid/4ed35bae2.html.

ICESCR—International Covenant on Economic, Social and Cultural Rights (1966). UN Treaty Collection. Optional Protocol to ICESCR Available: http://treaties.un.org/Pages/ViewDetails.aspx?src=TREATY&mtdsg_no=IV-3-a&chapter=4&lang=en [Accessed 22 April 2011.

UN General Assembly, International Covenant on Economic, Social and Cultural Rights, 16 December 1966, United Nations, Treaty Series, Vol. 993.

UN Committee on Economic, Social and Cultural Rights (CESCR), General Comment No. 3: The Nature of States Parties' Obligations (Art. 2, Para. 1, of the Covenant), 14 December 1990, E/1991/23.

UN Committee on Economic, Social and Cultural Rights (CmtESCR), General Comment No. 9, The domestic application of the Covenant (Nineteenth session, 1998), U.N. Doc. E/C.12/1998/24 (1998).

UN Committee on Economic, Social and Cultural Rights (CmtESCR), General Comment No. 15: The Right to Water (Arts. 11 and 12 of the Covenant), 20 January 2003.

UN Committee on Economic, Social and Cultural Rights (CESCR), General Comment No. 20: Non-discrimination in economic, social and cultural rights (art. 2, para. 2, of the International Covenant on Economic, Social and Cultural Rights), 2 July 2009, E/C.12/GC/20.

Vienna Decleration Programme of Action. 1993 June. *In*: World Conference on Human Rights (Vol. 25).

Reports and Resolutions

Australia: National Human Rights Consultation Report. September 2009. Chapter 15.

Resolution A/RES/64/292. The human right to water and sanitation. United Nations General Assembly, July 2010.

UN Secretary-General. 2002. Report on Activities undertaken in preparation for the International Year of Freshwater 2003, UN Doc. A/57/132.

World Health Organization and UNICEF Joint Monitoring Programme (JMP). 2015. Progress on Drinking Water and Sanitation, 2015 Update and MDG Assessment.

Case Law

Constitution of the Republic of South Africa Amendment Act. 1999. Section 27 (2).

Manqele v Durban Transitional Metropolitan Council in South Africa, Durban High Court, 2002 (6) SA 423 (D).

Mazibuko, L., Munyai, G., Makoatsane, J., Malekutu, S. and Paki, V. City of Johannesburg and Others (CCT 39/09) [2009] ZACC 28; 2010 (3) BCLR 239 (CC); 2010 (4) SA 1 (CC) (8 October 2009).

Paschim Banga Khet Mazdoor Samity v. State of West Bengal, (1996) AIR SC 2426/(1996) 4 SCC 37.

Soobramoney vs. Minister of Health, KwaZulu-Natal 1998 (1) SA 765 (CC), 1997 (12) BCLR 1696 (CC).

Articles & Books

Alston, P. and Goodman, R. 2012. International Human Rights. Oxford University Press.

Amnesty International (17th November 2016). Slovenia: Constitutional right to water "must flow down to" Roma communities. Available: https://www.amnesty.org/en/latest/news/2016/11/slovenia-constitutional-right-to-water-must-flow-down-to-roma-communities/. Last accessed 20/12/2016.

Barlow, M. 2008. Our Water Commons: Toward a new freshwater narrative. Council of Canadians.

Barlow, M. 2015. Our right to water. Assessing progress five years after the UN recognition of the Human rights to Water and Sanitation. Council of Canadians. Accessed online at (28/12/2016) Available at: [https://canadians.org/sites/default/files/publications/report-rtw-5yr-1115.pdf].

Bulto, T.S. 2011. The emergence of the human right to water in international rights law: Invention or discovery? *Melbourne Journal of International Law*, 12(2).

Coy, M., Lovett, J. and Kelly, L. 2008 . London: EVAW. Available at: http://www.endviolenceagainstwomen.org.uk/data/files/realising_rights.pdf.

Cullet, P. 2013. Right to water in India–plugging conceptual and practical gaps. *The International Journal of Human Rights*, 17(1).

Davidson-Harden, A., Bakker, K., Spronk, S. and McDonald, D. 2011. Local Control and Management of Our Water Commons. Stories of Rising to the Challenge.

Dennis, M. and Stewart, D. 2004. Justiciability of Economic, Social, and Cultural Rights: Should There Be an International Complaints Mechanism to Adjudicate the Rights to Food, Water, Housing, and Health? *The American Journal of International Law* 98(3): 462–515.

De Albuquerque, C. and Roaf, V. 2012. On the right track. Good practices in realizing the rights to water and sanitation. World Water Council, Lisbon. Accessed [01.01.2017]. Available at: http://www.ohchr.org/Documents/Issues/Water/BookonGoodPractices_en.pdf.

Donnelly, J. 2003. Universal Human Rights in theory and in Practice.

Drèze, J. 2004. Democracy and right to food. Economic and Political Weekly, pp. 1723–1731. https://casi.sas.upenn.edu/sites/default/files/iit/Democracy%20%26%20the%20Right%20 to%20Food%2C%20Jean%20Dreze%20-%20EPW.pdf.

Gauri, V. 2009. Public interest litigation in India: overreaching or underachieving? *World Bank Policy Research Working Paper Series, Vol.*

Gleick, P.H. 1998. The human right to water. *Water Policy* 1(5): 487–503.

Kothari, J. 2006. The Right to Water: A Constitutional Perspective. Paper prepared for the workshop, "Water, Law and the Commons", Delhi, 8–10 Dec 2004 organized by the International Environmental Law Research Centre.

Kelley, D. 1998. A life of one's own: Individual rights and the welfare state. Cato Institute.

Langford, M. 2008. Social Rights Jurisprudence: Emerging Trends in International and Comparative Law. Cambridge University Press.

Mehta, L., Allouche, J., Nicol, A. and Walnycki, A. 2014. Global environmental justice and the right to water: the case of peri-urban Cochabamba and Delhi. *Geoforum* 54: 158–166.

Meshel, T. 2015. Human Rights in Investor-State Arbitration: The Human Right to Water and Beyond. *Journal of International Dispute Settlement*, p.idv007.

Neier, Aryeh. 2006. Social and economic rights: a critique. Human Rights Brief 13, no. 2: 1–3.

Radonic, Lucero. 2015. Through the aqueduct and the courts: An analysis of the universal right to water and Indigenous water rights in Northwestern Mexico. Geoforum 84(1): 151–157.

Rawls, J. 2001. Justice as Fairness: A Restatement. Harvard University Press.

Rosenberg, G. 2008. 2nd edition. The Hollow Hope: Can Courts Bring about Social Change.

Salman, S.M. and McInerney-Lankford, S. 2004. Legal and Policy Dimensions.

Sen. 2009. The Idea of Justice. Harvard University Press.

Sen, A. 2011. The Idea of Justice. Harvard University Press.

Shue, Henry. 1996. Basic rights: Subsistence, affluence, and US foreign policy. Princeton University Press.

Swyngedouw, E., Castro, J.E. (ed.) and Kohan, G. (ed.). 2014. The hydro-social circulation of water. *In: Territorialidades del Agua: Conocimiento y Acción para Construir el Futuro que Queremos.*

Tully, S. 2005. Human right to access water—A critique of general comment no. 15, A. *Neth. Q. Hum. Rts.*, 23.

Tushnet, M. 2008. Weak Courts, Strong Rights: Judicial Review and Social Welfare Rights in Comparative Constitutional Law.

Winkler, I. 2008. Judicial Enforcement of the Human Right to Water—Case Law from South Africa, Argentina and India. *Law, Social Justice & Global Development Journal (LGD).*

Young, K. 2008. The minimum core of economic and social rights: a concept in search of content. *Yale International Law Journal* 33: 113–175.

Section 4
Water Security

Chapter-11

Water Security

Emerging Paradigm, Challenges and Opportunities

Velma I. Grover

INTRODUCTION

The world population is expected to reach the 9 billion mark by 2050 and water use is expected to increase 55% in the world water use between 2000 and 2050. Increase in urbanization, and an increase in demand for food and energy, also adds extra pressure on the environment and natural resources such as water, and exacerbate water risks. It is also projected that more than 40% of the global population will be living in severe water stress river basins—which means that more than 2.8 billion people will be living in water stress areas[1] by 2025, and the number is expected to rise to 4 billion by 2050.[2,3]

In addition, climate change and degradation of water quality increase the uncertainty of water availability.[4,5] Change in the climate affects the hydrologic

This is based on the publication: Water Security. Encyclopedia of Environmental Management DOI: 10.1081/E-EEM-120051581, York University, Toronto, Ontario, Canada.

[1] "Hydrologists typically assess scarcity by looking at the population–water equation. An area is experiencing water stress when annual water supplies drop below 1,700 m³ per person. When annual water supplies drop below 1,000 m³ per person, the population faces water scarcity, and below 500 cubic metres 'absolute scarcity'" (http://www.un.org/waterforlifedecade/scarcity.shtml).

[2] Gardner-Outlaw, T. and Engleman, R. 1997. Sustaining Water, Easing Scarcity: A Second Update; Population Action International: Washington, DC.

[3] UNFPA. 1997. Population and Sustainable Development: Five Years After Rio; UNFPA: New York.

[4] OECD. 2013. Water Security for Better Lives, OECD Studies on Water; OECD: Paris, 13.

[5] Planet Under Pressure. Water Security for a Planet Under Pressure (Rio +20 Policy Brief). In The international conference Planet Under Pressure: New Knowledge Towards Solutions, London, 2012. Available at www.planetunderpressure2012. net (accessed November 2013).

cycle[6,7]. As discussed in the Intergovernmental Panel on Climate Change report, water is part of all components of the climate system, that is, atmosphere, hydrosphere, cryosphere, land surface, and biosphere. Essentially, this means that any change in the climate impacts the water (cycle) through different means. Climate change has been associated with changes in the hydrological systems, such as change in precipitation patterns, melting of snow and ice, increased evaporation, increased atmospheric water vapour, and changes in soil moisture and run off.[8] As the temperature increases, saturation of vapour pressure of air increases; and it is expected that with global warming (increasing temperature), the amount of water vapour suspended in the air will increase.[9,10]

Both developing and developed countries face slightly different water security challenges. Non-OECD (Organization for Economic Cooperation for Development) countries will have a greater rate of population growth (and in some countries, such as India, the rate of income will exceed that of the OECD average) impacting water use and availability ratio. Some of the water challenges that the OECD countries face are as follows: the move towards water pricing based on supply costs together with water recycling investments and improvements in water use efficiency in agriculture have resulted in decoupling water demand from gross domestic product, the projection is that water demand will decrease in the OECD countries! As compared to non-OECD countries, OECD countries have better infrastructure for water services, water storage capacity per capita and share of hydropower potential, and comparatively more predictable and moderate rainfall, making them more resilient and prepared for disasters. For most of the developing countries, water security challenges include increasing water use and nutrient pollution; and the developed countries face water security challenges such as local water shortages, growing risk from floods, and financing to replace the aging infrastructure.[11]

This chapter begins with definitions of water security followed by a description of the key dimensions of water security. The next section describes the key indicators for measuring water security and the subsequent section discusses the main challenges and threats to water security. The following section details how water security can be achieved. The water security debate would be incomplete without looking at how it fits within the national security debate—the next section addresses this, followed by a comparison of water security approach with other water management paradigms.

---

[6] Alavian, V., Qaddumi, H.M., Dickson, E., Diez, S.M., Danilenko, A.V., Hirji, R.F., Puz, G., Pizarro, C., Jacobsen, M. and Blankespoor, B. 2009. Water and Climate Change: Understanding the Risks and Making Climate-Smart Investment Decisions. World Bank: Washington, DC. Available at http://documents.worldbank.org/curated/en/2009/11/11717870/water-climate change-understanding-risks-making-climate-smart-investment-decisions (accessed November 2013).

[7] Beddington, J. 2013. Catalysing sustainable water security: Role of science, innovation and partnerships. *Philosophical Transactions of the Royal Society A* 371: 20120414.

[8] Bates, B.C., Kundzewicz, Z.W., Wu, S. and Palutikof, J.P. (eds.). 2008. Climate Change and Water. Technical Paper of the Intergovernmental Panel on Climate Change; IPCC Secretariat: Geneva, 210 pp.

[9] Stocker, T.F., Qin, D., Plattner, G.-K., Tignor, M., Allen, S.K., Boschung, J., Nauels, A., Xia, Y., Bex, V., Midgley, P.M. and IPCC. 2013. Climate Change 2013: The Physical Science Basis. Contribution of Working Group I to the Fifth Assessment Report of the Intergovernmental Panel on Climate Change; Cambridge University Press: Cambridge, UK and New York, USA.

[10] Grover, V.I. (ed.). 2012. Impact of Climate Change on Water and Health. CRC: Boca Raton, FL.

[11] OECD. 2013. Water Security for Better Lives, OECD Studies on Water; OECD: Paris, 13.

What is Water Security, Some Definitions

Although there is an increased concern about water issues and water security, there is no clear, common definition of water security. Different authors from different disciplines and institutions have defined it in a somewhat different manner, different fashion, for example, a lawyer would frame the water security with water allocation rules, whereas an agriculture professional would look at the water security from the lens of flood and drought protection.[12]

"Traditionally, water security had two meanings which applied to the rights of an individual and to the claims of a state on behalf of its citizen."[13] In this case, the concept of water security evolved around a physically dependable supply, which is linked to a legal allocation of water share for a country/community.[14] The definition and concept of water security has evolved since the 1990s when it was linked to specific human security issues such as food security or military security. The Global Water Partnership, in 2000, came up with an integrative/holistic definition of water security that includes access, affordability of water, and human needs as well as ecological health. Since then a few other definitions have come up centred around four thematic areas: water availability; other human needs, such as food security; vulnerability of humans to hazards; and sustainability issues.[15,16] Definitions of water security also vary from one region to the other, depending on the local challenges and issues in the water sector. For example, in Australia, most of the water security debates are centred on water availability; but in China, the challenge is a growing demand for water and water pollution from industries leading to water security research, with a combined focus on both water availability and water pollution. On the other hand, the Middle East faces the challenge of sharing limited water resources; hence, water security focuses on sharing resources in an unstable geopolitical environment.[17]

Some definitions of water security are

Reliable access to potable water of sufficient quantity and quality for basic human needs, small-scale livelihoods and local ecosystem services, are coupled with a well-managed risk of water-related disasters.[18]

---

[12] Lankford, B., Bakker, K., Zeitoun, M. and Conway, D. 2013. Water Security: Principles, Perspectives and Practices. Routledge: New York.

[13] Salman, M.A. and Kishor Uprety. Shared watercourses and water security in south asia: challenges of negotiating and enforcing treaties. *International Water Law.* Vol 3(3).

[14] Salman, M.A. and Kishor Uprety. Shared watercourses and water security in South Asia: Challenges of negotiating and enforcing treaties. *International Water Law.* Vol 3(3).

[15] Dunn, G., Cook, C., Bakker, K. and Allen, D. (eds.). 2012. Defining and assessing water security (Part 1, Section 1). In Water Security Guidance Document, 8: 13. Available at http://watergovernance.ca/wp-content/uploads/2011/12/PART-1-SECTION-1.pdf (accessed November 2013).

[16] Cook, C. and Bakker, K. 2012. Water security: Debating an emerging paradigm. *Global Enviromental Change* 22: 94–102.

[17] Dunn, G., Cook, C., Bakker, K. and Allen, D. (eds.). 2012. Defining and assessing water security (Part 1, Section 1). In Water Security Guidance Document, 8: 13. Available at http://watergovernance.ca/wp-content/uploads/2011/12/PART-1-SECTION-1.pdf (accessed November 2013).

[18] Water Aid. 2012. Water Security Framework; Water Aid: London, 5: 48.

[Water security is] the availability of an acceptable quantity and quality of water for health, livelihoods, ecosystems and production, coupled with an acceptable level of water-related risks to people, environments and economies.[19]

Water security "means ensuring that freshwater, coastal and related ecosystems are protected and improved; that sustainable development and political stability are promoted, that every person has access to enough safe water at an affordable cost to lead a healthy and productive life, and that the vulnerable are protected from the risks of water-related hazards" (Ministerial Declaration of the 2nd World Water Forum).[20]

A comprehensive definition [of water security] goes beyond availability to issues of access. Access involves issues that range from a discussion of fundamental individual rights to national sovereignty rights over water. It also involves equity and affordability, besides the roles of states and markets in water's allocation, pricing, distribution and regulation. Water security also implies social and political decision-making on use—the priority to be accorded to competing household, agricultural or industrial demands on the resource.

Eric Guttierez and Patricia Wouters,[21]

The greatest water problem [is] our inability to link environmental security, water security and food security. Water security is linked to a safe water supply and sanitation, water for food production, hydro-solidarity between those living upstream and those living downstream in a river basin and water pollution avoidance so that the water in aquifers and rivers remain usable, i.e., not too polluted for use for water supply, industrial production, agricultural use or the protection of biodiversity, wetlands and aquatic ecosystems in rivers and coastal waters.

Malin Falkenmark,[22]

Social and physical processes combine to create or deny water security. Sustainable water security is interpreted as a function of the degree of equitability and balance between interdependencies of the related security areas, plated out within a web of socioeconomic and political forces at multiple spatial levels—the "web" of water security identifies the "security areas" related to national water security. These include intimately associated natural "security resources" (water resources, energy, climate, food) as well as the security of the social groups concerned (individual, community, nation). The "web" recognizes the interaction occurring at all spatial scales, from the individual through to river basin and global levels. In this sense, an individual's water security may coexist with national water insecurity, as in the case of wealthy farmer-sheikhs with the deepest wells (who maybe temporarily water-secure) in the dry highlands of Yemen (which is not, water secure).

Mark Zeitoum[23]

---

[19]   Grey, D. and Sadoff, C.W. 2007. Sink or swim? Water security for growth and development. *Water Policy* 9: 545–571.

[20]   Planet Under Pressure. 2012. Water Security for a Planet Under Pressure (Rio +20 Policy Brief). In The international conference Planet Under Pressure: New Knowledge Towards Solutions, London. Available at www.planetunderpressure2012. net (accessed November 2013).

[21]   Water Aid. 2012. Water Security Framework; Water Aid: London, 5: 48.

[22]   Water Aid. 2012. Water Security Framework; Water Aid: London, 5: 48.

[23]   Water Aid. 2012. Water Security Framework; Water Aid: London, 5: 48.

The gossamer that links together the web of food, energy, climate, economic growth and human security challenges that the world economy faces over the next two decades.[24]

Sustainable access on a watershed basis to adequate quantities of water, of acceptable quality, to ensure human and ecosystem health.[25]

Water security involves the sustainable use and protection of water systems, the protection against water related hazards (floods and droughts), the sustainable development of water resources and the safeguarding of (access to) water functions and services for humans and the environment.[26]

There are three important elements of "water security":

1. Water security is based on three core freedoms: freedom from want, freedom from fear and freedom to live with human dignity;

2. Ensuring water security may lead to a conflict of interests, which must be identifiable and effectively dealt with at the international, national and local levels;

3. Water security, like water, is a dynamic concept, and one that needs clear local championship and sustained stewardship.

Patricia Wouters[27]

Water security is the availability of freshwater in the right quantity and quality, at the right times, for dependent systems.[28]

Water security is about managing water risks, including risks of water shortages, excessive pollution, and risks of undermining the resilience of freshwater systems.[29]

Key Dimensions of Water Security

According to Asian Development Bank (ADB), there are five key dimensions of water security, which can also be linked to five Maslow Hierarchical Needs[30] of Maslow. As the Maslow Hierarchy of needs mention that the lower levels of needs such as food, water and air security need to be met, followed by home, employment safety, and feeling of belongingness before moving to higher levels of self esteem and self actualization, similarly in the water security pyramid the key to achieve

[24] World Economic Forum (WEF). 2009. The Bubble is Close to Bursting: A Forecast of the Main Economic and Geopolitical Water Issues Likely to Arise. World Economic Forum: Davos, 5.

[25] Dunn, G., Cook, C., Bakker, K. and Allen, D. (eds.). 2012. Defining and assessing water security (Part 1, Section 1). In Water Security Guidance Document, 8: 13. Available at http://watergovernance.ca/wp-content/uploads/2011/12/PART-1-SECTION-1.pdf (accessed November 2013).

[26] Schultz, B. and Uhlenbrook, S. 2007. Water Security: What Does It Mean, What May It Imply; Discussion draft paper, UNESCO-IHE Institute for Water Education: Delft, The Netherlands, 2.

[27] Water Aid. 2012. Water Security Framework; Water Aid: London, 5: 48.

[28] Petersen-Perlman, J.D., Veilleux, J.C., Zentner, M. and Wolf, A.T. 2012. Case studies on water security: Analysis of system complexity and the role of institutions. *Journal of Contemporary Water Research & Education* 149: 4–12. doi:10.1111/j.1936-704X.2012.03122.x.

[29] OECD. 2013. Water Security for Better Lives, OECD Studies on Water; OECD: Paris, 13.

[30] Asian Development Bank. 2013. Asian Water Development Outlook 2013: Measuring Water Security in Asia and the Pacific; Asian Development Bank: Mandaluyong City, The Philippines.

highest level of water security (which is security at a national level and the need to build water-related disaster resilient communities), efforts need to start from the bottom most level: household levels. Every household should have access to safe drinking water and sanitation followed by balancing water needs for economy (i.e., agriculture, industry, and energy needs of the communities). Household water security level is followed by urban water security—which goes beyond mere access to safe drinking water to creating livable cities by providing wastewater treatment facilities and drainage. The next step up in the water security pyramid is environmental water security achieved by building policies, management strategies, and local capacities for disaster-resilient communities. Similar to the Maslow hierarchy of needs where first the lower level of needs should be met to achieve the higher level of needs—the pyramid of water security follows the same principles where household-level water security needs to be achieved before national water security issues can be dealt with.

Methods to Measure Water Security

For a policy maker (and others), it is important to know the water security of an area which involves measuring water security/insecurity. Usually, water security is measured in terms of water access or water availability, but there is no one easy way to measure water security/insecurity. There are a few different indicators or methods to measure water security such as Falkenmark Water Stress Indicator, Water Poverty Index, Mapping Water Scarcity, Water Security Status Indicators, ADB indicators, Overseas Development Institute indicators, etc. A couple of these are discussed below:

ADB indicators

Linked to the key dimensions of water security discussed above, the National Water Security index gives progress of countries toward meeting national water security and combines information of the five key elements discussed subsequently:[31]

- Household water security satisfies household water and sanitation needs in all communities. Meeting household water security works as a strong foundation to support development, to eradicate poverty, and to meet national water security. This index gives an indication of the current status (and progress made) by countries in meeting household water and sanitation needs (including hygiene)—by measuring percentage access to piped water supply, improved sanitation, and hygiene.

- Economic water security supports productive economies in agriculture and industry. This is measured and the productive use of water that can sustain the economic growth in all the sectors—agriculture, industry, and energy.

---

[31]   Asian Development Bank. 2013. Asian Water Development Outlook 2013: Measuring Water Security in Asia and the Pacific; Asian Development Bank: Mandaluyong City, The Philippines.

- Urban water security measures progress toward the creation of better water management services that support the development of vibrant, livable cities and towns.

- Environmental water security gives an indication on how well river basins are developed and managed to sustain ecosystem services.

- Water-related disaster resilience measures the capacity of communities to cope with and recover from water related disasters. It is essential to go beyond providing basic water and sanitation services to actually building resilience of communities.

Lautze and Manthrithilake[32]

This index looks at the country-level water security. The index considers five key components of water security (basic needs, agricultural production, environment, risk management, and independence; see Table 11.1) and translates them into numerical indicators. Each component is placed on a 5-point scale, which is an indication of the degree of water security for that water component (for example: for household level the scale numbers the percentage of population with sustainable access to an improved water source from 1 to 5), and a total of 25-point scale is a sum of all the five components and gives an indication of the degree of water security for a country. One of the fundamental issues addressed by this indicator is the difference between relative versus absolute water security. As mentioned by the authors, "Interpretations of water security could benefit from clear focus on the end of water security—not the means to water security, and not the ends beyond water security."

Table 11.1 Water security components.

Overall Water Security = A + B + C + D + E
Component Definition
A = Basic household needs: The percentage of population with sustainable access to an improved water source
B = Agricultural production: The extent to which water is available and harnessed for agricultural production
C = Environmental flows: Percentage of renewable water resources available for environmental water requirement
D = Risk management: The extent to which countries are buffered from the effects of rainfall variability through large dam storage
E = Independence: The extent to which countries' water and food supplies are safe and secure from external changes or shocks

Source: Adapted from Lautze & Manthrithilake.[33]

[32] Lautze, J. and Manthrithilake, H. 2012. Water Security: Old Concepts, New Package, What Value? GWF Discussion Paper 1250; Global Water Forum: Canberra, Australia.

[33] Lautze, J. and Manthrithilake, H. 2012. Water Security: Old Concepts, New Package, What Value? GWF Discussion Paper 1250; Global Water Forum: Canberra, Australia.

Challenges and Threats to Water Security

Usually people are not water insecure because there is a lack of water availability, but because of poor governance; lack of investment in water infrastructure and management; insufficient skills or human capital to manage water supply; lack of political will; and inability of people to pay.[34]

Seven major challenges to achieving water security were identified by the Ministerial Declaration at the Second World Water Forum in 2000: meeting basic needs, securing the food supply, protecting ecosystems, sharing water resources, managing risks, valuing water, and governing water wisely.[35] Some of these challenges and threats to water security are discussed in detail below.[36]

Poor Governance and lack of (or Weak) political will

If the government will is lacking, there is not enough allocation of funds for infrastructure and institutional capacity building or even in development of human capacity. Even if there is an allocation of funds, there is no human capacity to implement the targets. Both urban and rural areas have slightly different problems. One of the major problems in urban areas is a leakage from the water supply system mainly because infrastructure is not properly managed. The main issue in rural water services are delegated to communities is that local partnership and ownership are necessary for successful water projects, most of the communities cannot manage it without external technical, managerial, and financial support.

Social and political exclusion

In some cases, the problem is not the availability of water, but the exclusion of some communities based on their social status, their political affiliation (or power asymmetry), race, gender, or age (in some countries, caste also plays a role). There can be no one size fits all but one of the solutions can be implementation of the UN General Assembly resolution (2010) for access to safe water and sanitation as a human right.

Economic exclusion

In some communities, the people who cannot pay are excluded from the system, even if that community has safe drinking water supply. Research conducted in Ethiopia shows a direct link between wealth, access, and water usage.

---

[34] Water Aid. 2012. Water Security Framework; Water Aid: London, 5: 48.

[35] Planet Under Pressure. 2012. Water Security for a Planet Under Pressure (Rio +20 Policy Brief). In The international conference Planet Under Pressure: New Knowledge Towards Solutions, London. Available at www.planetunderpressure2012. net (accessed November 2013).

[36] Water Aid. 2012. Water Security Framework; Water Aid: London, 5: 48.

Low resilience to cope with stresses on water supplies

Communities that have fragile ecosystems, vulnerable livelihoods, and limited financial resources also have limited technical and adaptive capacity to deal with water stress issues, making them more water insecure.

Poor hygiene and lack of sanitation facilities

Even if clean water is supplied, poor hygienic habits of people might lead to issues such as storing water in dirty collection vessels and using unwashed hands in handling water for drinking as well as cooking. Also, the lack of sanitation facilities creates an environment for pathogens to contaminate clean drinking water. All of these factors decrease water quality and increase the risks of water related diseases, making the area/community water insecure.

Rapid population growth and urbanization pose certain threats to water security

Rapid population growth and urbanization pose threats such as: increased domestic demand for water; increased pressure on already struggling governments and service providers to keep pace with increasing water demand in both urban and rural areas; increased demand for agriculture (and meeting food security); increased competition for water among different users and countries (in case of transboundary water); risks of water pollution from intensive agriculture and heavy industrial use; and increased disaster risk (especially for people settled on marginal lands).

Climate variability and climate change

Climate variability is one of the greatest challenges for water (and food) security because it is changing the weather patterns and the timings and amount of precipitation (causing floods or droughts in some places, impacting agriculture). This has a great impact on people dependent on rain for agriculture and their livelihood. Climate change is an important long-term risk, and although there are a lot of uncertainties surrounding it, action must be taken to adapt to/mitigate these changes.

Growing demand for food

Food security is very closely linked to water security. Water scarcity affects food production which can cause a shortage of even water . Both water and food scarcity lead to malnutrition and related health problems. With the changing food basket and increasing number of people consuming meat, eggs, and dairy, livestock production has increased and it has also impacted the world's water supply, almost accounting for 8% (or more) of global human water use (mainly for growing feed crops).[37]

---

[37] FAO. 2006. Livestocks Long Shadow; FAO: Rome.

Another factor impacting animal production will be extended period of droughts, which will cause shortages of food and drinking water.[38]

Different Approaches to Meet Water Security Challenges

To address the complex 21st century water security challenges, innovative solutions and strategies are needed including: ensuring efficient water-related services for 9 billion people; managing water-related threats to society; establishing global architecture to work across disciplines and science-based policy divides; filling the water knowledge gap and ensuring knowledge flow from local to global levels; and improving cooperation between interdisciplinary and international water scientists and policy makers.[39]

As mentioned above in the introductory section, perspective on water security varies from region to region depending on the local issues—which means the solutions to meet water security will be local as well. For example, in South Asia the main problem is that of water scarcity, and as proposed by a panel discussion on water security,[40] the solution to this issue lie in three "i's,": infrastructure (for water storage, development of hydropower, and water supply), institutions (robust institutions and a good governance system to deal with challenges), and innovations (inclusive of local cultural and religious issues). Some other solutions to deal with water security include:

- The role of nature can also be explored more to achieve water security such as: allocation of water for ecological and ecosystem services,[41] and payment for ecosystem services.[42]

- Technical (traditional scientific) methods or techniques, such as the Zai technique (it captures rain and surface run-off water to be used in agriculture),[43] ecological engineering, and reforestation.

- However, the focus of the following discussion for the solutions to deal with water security challenges is more on the policy level. Although some authors fear

---

[38] Sundstrom, J., Albihn, A., Boqvist, S., Ljungvall, K., Marstrop, H., Martiin, C., Nyberg, K., Vagsholm, I., Yuen, J. and Magnusson, U. 2014. Future Threats to Agricultural Food Production Posed by Environmental Degradation, Climate Change and Animals and Plant Diseases—A Risk Analysis in Three Economic and Climate Settings; Food Security; Springer: Netherlands, Feb. 2014.

[39] Grey, D., Garrick, D., Blackmore, D., Kelman, J., Muller, M. and Sadoff, C. 2013. Water security in one blue planet: twenty-first century policy challenges for science. *Philosophical Transactions of the Royal Society A* 371: 4.

[40] Briscoe, J., Babar Ali, S., Eck, D., Henderson, R. and Yu, W. 2011. The future of South Asia Conference, The South Asia Initiative at Harvard University (Panel Talk on Water Security), Harvard University, Cambridge, Apr 9, 2011.

[41] FAO. 2000. New Dimensions in Water Security: Water, Society and Ecosystem Services in the 21st Century; FAO: Rome.

[42] Sundstrom, J., Albihn, A., Boqvist, S., Ljungvall, K., Marstrop, H., Martiin, C., Nyberg, K., Vagsholm, I., Yuen, J. and Magnusson, U. 2014. Future Threats to Agricultural Food Production Posed by Environmental Degradation, Climate Change and Animals and Plant Diseases—A Risk Analysis in Three Economic and Climate Settings; Food Security; Springer: Netherlands, Feb. 2014.

[43] Burkina Faso: The Zaï technique and enhanced agricultural productivity. May 2005. Available at http://www.worldbank.org/afr/ik/iknt80.htm (accessed July 2014).

that the word "security" carries some kind of military undertone or connotation and "water security" can be achieved only by force instead of negotiations,[44,45] some of the general (non-military) ways to approaching water security are discussed below.

ADB approach of meeting water security

According to the framework developed by ADB, water security is enjoyed when water is managed successfully and the following services can be implemented: meet household water and sanitation needs at community levels; support various economic stakeholders, that is, agriculture, industry, and energy; be able to meet demands in the urban cities to maintain livable cities and towns; restore and maintain rivers and ecosystems; and build resilient cities (the hierarchical structure discussed above).[46] To achieve water security, it is important that governments look at the interdependence of all the five key elements/dimensions of water security and have a governance/institutional system that can go beyond the traditional sector silo approach to more integrated and holistic approach.

Policy interventions required to meet each key dimension are discussed below (based on discussion in the Asian Development Bank):[47] For household water security four strategic interventions required include: financial, management, social, and technical. Governments should integrate financing for water supply and sanitation into their national budgeting system. To make water supply and sanitation services inclusive of all people, there should be an effort to internalize connection costs of water supply into tariffs or long-term installment payment plans should be offered. The second priority should be for a transparent governance system with a focus on performance management systems, including, monitoring and assessment. Government policies should promote water conservation methods either through water-saving household technologies, regulations for industry, and development/use of recycling and reuse technologies.

Community-managed rural water supply systems and community-led total sanitation programs (including support to zero open defecation programs) should be promoted to provide local ownership, external technical and financial support should be provided to move them ahead.

Water is critical to support productive economies in the agriculture, industry, and energy sector, which in turn provide employment and food, and improve socioeconomic conditions of people/communities. There are three different options to achieve economic water security: continue the expansion of water supply by increasing water storage infrastructures to reduce variability in supply; improve

---

[44] Mason, N. and Calow, R. 2012. Water Security: From Abstract Concept to Meaningful Metrics: An Initial Overview of Options; Overseas Development Institute: London, 26.

[45] Mason, N. and Calow, R. 2012. Water Security: From Abstract Concept to Meaningful Metrics: An Initial Overview of Options; Overseas Development Institute: London, 26.

[46] Asian Development Bank. 2013. Asian Water Development Outlook 2013: Measuring Water Security in Asia and the Pacific; Asian Development Bank: Mandaluyong City, The Philippines.

[47] Asian Development Bank. 2013. Asian Water Development Outlook 2013: Measuring Water Security in Asia and the Pacific; Asian Development Bank: Mandaluyong City, The Philippines.

the productivity; and transform the national economic mix by promoting activities that would better match the available water resources (using water for the most optimal use). Policy levers to improve economic water security include: financing infrastructure to support new policies; governance (improving governance to support trans-boundary institutional framework, basin-wide water allocation systems, and explore the role of private sector in irrigation); investment in irrigation and rain-fed agricultural production; promote irrigation technologies to improve water productivity and promote agricultural trade between water-abundant and water scarce regions; improve energy use in industries and encourage reuse of water in the industrial sector; encourage use of alternate forms of renewable energy; and minimize use of water consumption in activities, such as thermal power generation.

The third dimension to support urban vibrant cities will involve changes in the institutional and public perceptions on water management. The water-sensitive cities framework helps in policy and investment planning for a transition of an urban area/city from providing basic services to increasingly value-added services. The idea behind the water-sensitive cities framework is that city managers are trying to improve water services to achieve improved water supplies, separated sewerage schemes, and also effective drainage and flood protection.[48] Policy levers to increase urban water security include investments in economically and financially feasible water and wastewater infrastructure to provide water and also protect rivers; strengthen river basin organizations and explore the potential for corporatizing utilities or including the private sector; raise public awareness to conserve water; invest in flood forecasting, environmental monitoring systems to build flood/disaster resilience; and formulate policies for both sustainable water supply and long-term environmentally sound condition of urban water sources. In some countries, for example, Tanzania, a lot of surface as well as groundwater is available. However, the problem is the lack of capacity of local governments and communities to access water. Lack of access to clean drinking water is not because of real water shortages but lack of technical and financial capacity to tap into the resource (http://blogs.worldbank.org/africacan/tanzania-water-is-life-but-access-remains-aproblem).

To achieve environmental water security, some of the policy levers include: approaches to rehabilitate and protect rivers, which will need investments for river cleanups, and improved waste water treatment technologies as well as infrastructures (in addition to policies, regulation, and implementation strategy). In this case, implementation of the supportive policies and regulations will be important and will need adequate investment for implementation.

Some of the other ways to achieve environmental water security include prompting payment for watershed services programs, encouraging behavioural changes by community awareness programs, and adopting more integrated and holistic water resources management approaches. In conclusion, to improve the management and protection of rivers and watersheds, a well-defined water rights

---

[48]   Asian Development Bank. 2013. Asian Water Development Outlook 2013: Measuring Water Security in Asia and the Pacific; Asian Development Bank: Mandaluyong City, The Philippines.

approach (which also encourages payment-for environmental services scheme) and effective water allocation systems are essential.

To build resilient communities, emphasis should be on community-based approaches and climate change adaptation measures (deeply incorporated into the national policy) to deal with the risk of water-related disasters. Policy levers to increase water-related disaster resilience include investments in disaster risk management and allocating funds for response (proportionate to the level of risk) and providing disaster damage insurances, for example, to farmers for crop loss; increase investments into early warning systems and raise awareness of communities on disaster preparedness and response measures (including incorporating some of this in school curricula); and investment in environmental and climate monitoring, forecasting systems. Economic development and resilience are co-related; better resilience is good for higher economic development, and it also decreases the risk of loss of critical infrastructure (and thus save reconstruction costs).

Applying a risk-based approach to water security (OECD)

Most of the definitions of water security look at the water access and how to deal with water disasters. Both can be viewed from the perspective of "societal risks", which would then include risks associated with variable water supply as well as unpredictable water-related events. Based on this discussion, another way of defining water security, "is a tolerable level of water-related risk to society".[49]

Lautze and Manthrithilake[50] have identified five components that are critical to achieving water security. Hall and Borgomeo[51] have taken them as indicator risks to water security: "(i) of not satisfying basic needs (for given proportions of time and of the population), (ii) of inadequate agricultural production owing to water-related constraints, (iii) of harmful environmental impacts, and (iv) to the reliability of water supplies from the actions of neighbouring countries. Thinking of water security as the absence of intolerable risks leads to consideration of a broad range of water-related risks and context-specific evaluation of their tolerability."

One of the intrinsic elements to water security is that the water source may deteriorate (in terms of either quality or quantity) and have an impact on human or ecological health.[52] [This has antecedents in the 2000 UN Ministerial Declaration (in The Hague) for water security in the 21st century. The Ministerial Declaration listed seven core challenges to achieve water security: (1) meeting basic human needs; (2) securing food supply and water population growth for food production;

---

[49] Grey, D., Garrick, D., Blackmore, D., Kelman, J., Muller, M. and Sadoff, C. 2013. Water security in one blue planet: twenty-first century policy challenges for science. *Philosophical Transactions of the Royal Society A* 371: 4.

[50] Lautze, J. and Manthrithilake, H. 2012. Water Security: Old Concepts, New Package, What Value? GWF Discussion Paper 1250; Global Water Forum: Canberra, Australia.

[51] Hall, J. and Borgomeo, E. 2013. Risk-based principles for defining and managing water security. *Phil. Trans. R Soc. A* 371: 20120407, 4.

[52] Dunn, G., Cook, C., Bakker, K. and Allen, D. (eds.). 2012. Defining and assessing water security (Part 1, Section 1). In Water Security Guidance Document, 8: 13. Available at http://watergovernance.ca/wp-content/uploads/2011/12/PART-1-SECTION-1.pdf (accessed November 2013).

(3) ensuring integrity of ecosystems; (4) sharing water resources and promoting peaceful cooperation; (5) managing risks; (6) valuing water to reflect its economic, social, environmental, and cultural values; and (7) ensuring good governance through stakeholder participation].[53]

To achieve water security, the following four types of water risks need to be maintained at acceptable risk levels:[54]

- Risk of shortage (including droughts): the aim of this risk is to assess whether enough water is available or there is a risk of insufficient water available to meet demand (both short-term and long-term demand) for (beneficial) use by all water users (including environment).

- Risk of inadequate quality: the concern here is to check whether there will be a lack of water of suitable quality for a particular use (e.g., drinking, agriculture, industry).

- Risk of excess (including floods): in this case, the risk is evaluated for an overflow of a water system (either natural or human made), or submerging of an area that is usually not submerged with water.

- Risk of undermining the resilience of freshwater systems: Under this category, assessment is done to check whether the coping capacity of the surface or groundwater bodies (and their interactions) might be exceeded or the tipping point is crossed (as it can cause irreversible damage to the system's hydraulic and the biological functions of the ecosystem).

It is important that all the four risks are assessed simultaneously because[55]

- they can impact each other due to their interlinkages through the water cycle; and

- management of all these risks is important to achieve water security.

Although ADB and OECD approaches are different, yet they are not mutually exclusive, there are some overlaps, for example, ADB also has some risk aspects such as disaster resilience.

Achieving Water Security Targets through Market-Based Instruments

Some of the ways to achieve water security management include supply management, demand management, quality improvement, and efficient and equitable allocation of water among users. Economic instruments can be used to address water supply, water

---

[53] Leb, C. and Wouters, P. 2013. The water security paradox and international law: Securitisation as an obstacle to achieving water security and the role of law in desecuritising the world's most precious resource. pp. 26–45. *In*: Bruce, K.B., Zeitoun, M. and Lankford, D.C. (eds.). Water Security: Principles, Perspectives, and Practices. Routledge: New York.

[54] OECD. 2013. Water Security for Better Lives, OECD Studies on Water; OECD: Paris, 13.

[55] OECD. 2013. Water Security for Better Lives, OECD Studies on Water; OECD: Paris, 13.

demand, water quantity and quality, but use of economic/market instruments is very context and situation specific, and one solution will not fit all. However, the two key principles of economic management of water that can be universally applied include efficiency (maximizing the welfare that is obtained from a resource by allocating it to the most valuable economic use) and equity (concerned with fairness of the allocation of resources across a given population).[56]

Trans-boundary water security using international law

ADB's Maslow Hierarchical comparison provides a framework to achieve water security up to the national level; however, a lot of water bodies are shared and transboundary water security needs to look at interaction between different levels: (1) global/regional (2) basin, and (3) individual/user.[57] For global/regional levels, as discussed by Wouters et al.[58] from International Law perspective and the 3-A Analytical Framework on water security, there are three core elements of water security: availability, access, and addressing conflicts of use.

Availability means that an adequate quality and quantity of water should be available at an adequate risk. Access means that water should be accessible to a broad range of stakeholders through enforceable rights to water. Addressing conflicts of use means: whenever there are conflicts related to water use either of inadequate quantity of water or access to transboundary water resources, these conflicts should be resolved through dispute settlement mechanisms.

International law provides a suite of rules and processes that guide states in addressing such water insecurities through de-securitization. For example, Ramsar treaty or UN Convention on Combating Desertification, or the Law of Nations can act as a platform for regional peace and security, or international water law (principles codified in the 1997 Convention of the Law of the non-navigational uses of international watercourse and United Nations Economic Commission for Europe (UNECE) convention on the protection and use of trans-boundary watercourses and international lakes).[59]

At the global level, treaties are more general, but at the basin level agreements can be more specific to the situation and needs of individual countries or states sharing the water body. This takes into account the risks associated with emergencies and natural hazards, and focus should also be on prompting water for human development and environmental needs.

---

[56] OECD. 2013. Water Security for Better Lives, OECD Studies on Water; OECD: Paris, 13.

[57] Leb, C. and Wouters, P. 2013. The water security paradox and international law: Securitisation as an obstacle to achieving water security and the role of law in desecuritising the world's most precious resource. pp. 26–45. *In*: Bruce, K.B., Zeitoun, M. and Lankford, D.C. (eds.). Water Security: Principles, Perspectives, and Practices. Routledge: New York.

[58] Wouters, P., Vinogradov, S. and Magsis, B.O. 2009. Water security, hydrosolidarity, and international law: A river runs through it. *Yearbook of International Environmental Law* 19: 97–134.

[59] Leb, C. and Wouters, P. 2013. The water security paradox and international law: Securitisation as an obstacle to achieving water security and the role of law in desecuritising the world's most precious resource. pp. 26–45. *In*: Bruce, K.B., Zeitoun, M. and Lankford, D.C. (eds.). Water Security: Principles, Perspectives, and Practices. Routledge: New York.

International law also supports individual water security. For example, approved human right to water and sanitation supports individual water rights, but there are other treaties and laws supporting individual water security (see Leb and Wouters).[60]

Linking Water Security to the National Security Debate

As defined by Stephen Walt,[61] security studies is "the study of the threat, use and control of military force." However, the emerging issues, such as health, sanitation, natural disasters, climate change, access to clean drinking water, and so on, challenge this definition of security by Walt.

Technically (and traditionally) human security has always been part of the "security" debate because "security traditionally has focused on the state because its fundamental purpose is to protect its citizens."[62] As states have not always been able to protect its citizens, there is a need to protect people, and the UN Commission of Human Security have correctly mentioned that human security means "protecting people from critical and pervasive threats and situations…creating systems that give people the building blocks of survival, dignity and livelihood."[63] The whole notion of human security revolves around protecting humans and giving them basic rights to live with dignity.

Essentially, "human security is about protecting people, providing peoples the opportunities for progress, promoting stability, security, sustainability."[64] In future, water management paradigms would be impacted by both politico-military and economic imperatives, and the water security debate will also be influenced by human security perspective. "Water poverty" is already defined in terms of human insecurity.[65] Also, human security in the context of water security can be interpreted as: all the individuals having access to clean and safe drinking water and sanitation (which will also prevent water-borne diseases).[66] As water is one of the most essential resources for human survival, in a way human security and water security are closely

---

[60] Leb, C. and Wouters, P. 2013. The water security paradox and international law: Securitisation as an obstacle to achieving water security and the role of law in desecuritising the world's most precious resource. pp. 26–45. *In*: Bruce, K.B., Zeitoun, M. and Lankford, D.C. (eds.). Water Security: Principles, Perspectives, and Practices. Routledge: New York.

[61] Walt, S.M. 1991. The renaissance of security studies. *International Studies Quarterly* 35(2): 212.

[62] Axworthy, L. 2001. Human security and global governance: Putting people first. *Global Governance* 7: 19.

[63] United Nations Commission on Human Security. Outline of the Report of the Commission on Humans Security. Available at http://www.unocha.org/humansecurity/chs/finalreport/Outlines/outline.pdf (accessed November 2013).

[64] Liotta, P.H., Mouat, D.A. and Lancaster, W.G. 2008. Environmental Change and Human Security: Recognizing and Acting on Hazard Impacts; Springer: Dodrecht, 2.

[65] Mason, N. and Calow, R. 2012. Water Security: From Abstract Concept to Meaningful Metrics: An Initial Overview of Options; Overseas Development Institute: London, 26.

[66] Ouyang, S.-Y., Zhao, T.Q., Wang, R.S., Leif, S. and Zhang, Q.X. 2004. Scenario simulation of water security in China. *Journal of Environmental Sciences* 16(5): 765–769.

linked.[67] In addition to human security, water security can be looked through the lens of "environmental security." Environmental security can be looked from two different perspectives—eco-centrism (the belief that all life, and the ecological systems on which life depends, possess intrinsic value)[68] or Neo-Malthusian (the environment with natural resources, focusing on the need to manage these resources in equitable and sustainable fashions),[69] but water security combines these two perspectives, because water security combines maintaining eco-integrity of the water ecosphere and watershed management aspects.[70] When water security is viewed more broadly, it encompasses food security (as agriculture is dependent on water), economic security (water is important for economic production activities, including energy and industry), environmental security (as water is important to preserve the environment), and human security (water is an essential resource for human survival, and safe drinking water is important for a healthy life). In one way, oil security is also dependent on water security because every step—discovery, extraction, and refining of oil—requires water.

Although there are a lot of discussions about water–food–energy nexus and some discussions in the trade arena because of "virtual water" concept, water security is still not part of foreign policies or defense community portfolios as climate security and resource scarcity.[71] However, some reports have specifically looked at the Global Water Security and its implications for the United States,[72,73] which concluded that "within 10 years water insecurity could be a contributing factor to state failure, and increasingly feature as a mechanism for contestation and leverage between states. Beyond 10 years, the report has high confidence that water is more likely to be used as a weapon by states or terrorists."[74] In addition to the trans-boundary water resources management, there are some other national security implications for water that look at the national water resources management and its economic development and stability.[75]

---

[67] Stomer, C. 2011. Water Security: Concepts. Challenges, and the South Sasketchewan River Basin; Master Thesis. University of Calgary: Calgary.

[68] Ishiyama, J.T. and Breuning, M. 2011. 21st Century Political Science: A Reference Handbook; SAGE Publications Inc.: Thousand Oaks, CA, 443.

[69] Ishiyama, J.T. and Breuning, M. 2011. 21st Century Political Science: A Reference Handbook; SAGE Publications Inc.: Thousand Oaks, CA, 443.

[70] Stomer, C. 2011. Water Security: Concepts. Challenges, and the South Sasketchewan River Basin; Master Thesis. University of Calgary: Calgary.

[71] Mason, N. and Calow, R. 2012. Water Security: From Abstract Concept to Meaningful Metrics: An Initial Overview of Options; Overseas Development Institute: London, 26.

[72] Mason, N. and Calow, R. 2012. Water Security: From Abstract Concept to Meaningful Metrics: An Initial Overview of Options; Overseas Development Institute: London, 26.

[73] NIC. 2012. Global Water Security—Intelligence Community Assessment; Office of the Director National Intelligence: Washington, DC.

[74] Mason, N. and Calow, R. 2012. Water Security: From Abstract Concept to Meaningful Metrics: An Initial Overview of Options; Overseas Development Institute: London, 26.

[75] Mason, N. and Calow, R. 2012. Water Security: From Abstract Concept to Meaningful Metrics: An Initial Overview of Options; Overseas Development Institute: London, 26.

Conclusion

Water security is an overarching conceptual framework that is inclusive of competing land and water-use practices and tries to balance them.[76] As observed in the discussion, there is no food, energy, environment, or human security without achieving water security—it is important to understand the concept and work toward achieving water security.

The World Bank has also mentioned that achieving water security is critical for growth and development.[77,78] Some countries, such as China,[79] have realized the importance of water security for sustainable development, and some other countries are using the concept for negotiations, for example, Cooperative Framework Agreement in the Nile Basin.[80,81] Just like there are so many definitions of water security, there are different indicators for measuring water security and achieving water security. With changing climate and changing hydrological cycle, water security has started featuring in national debate (e.g., the Global Water Security study carried out by the United States).

The best way to conclude is with the quote:

Water sits at the nexus of so many global issues—including health, hunger and economic growth. And sadly, water scarcity takes its greatest toll on the society's least fortunate. I am absolutely convinced that the only way to measurably and sustainably improve this dire situation is through broad-scale collaborative efforts between government, industry, academia and other stakeholders around the world.

Indra Nooyi, p.55[82]

---

[76] Dunn, G., Cook, C., Bakker, K. and Allen, D. (eds.). 2012. Defining and assessing water security (Part 1, Section 1). In Water Security Guidance Document, 8: 13. Available at http://watergovernance.ca/wp-content/uploads/2011/12/PART-1-SECTION-1.pdf (accessed November 2013).

[77] Lautze, J. and Manthrithilake, H. 2012. Water Security: Old Concepts, New Package, What Value? GWF Discussion Paper 1250; Global Water Forum: Canberra, Australia.

[78] Grey, D. and Sadoff, C.W. 2007. Sink or swim? Water security for growth and development. *Water Policy* 9: 545–571.

[79] Liu, B., Mei, X., Li, Y. and Yang, Y. 2007. The connotation and extension of agricultural water resources security. *Agricultural Sciences in China* 6(1): 11–16.

[80] Lautze, J. and Manthrithilake, H. 2012. Water Security: Old Concepts, New Package, What Value? GWF Discussion Paper 1250; Global Water Forum: Canberra, Australia.

[81] Water Link. 2010. Nile Basin Initiative Deadlock; Water Link International: Netherlands, July 6, 2010.

[82] Waughray, D. (ed.). 2011. Water Security: The Water-Food-Energy-Climate Nexus (World Economic Forum Water Initiative); Island Press: Washington, DC, 5. Water Security 15.

Chapter-12

Re-Defining Water Security

Amani Alfarra[1,*] and *Islam Abdelgadir*[2]

INTRODUCTION

This chapter focuses on redefining water security taking in consideration the political instability. Water security is defined as "access to safe and sufficient water for domestic use and sanitation". For this chapter, the human-centric approach is used which takes the household as the level of analysis. The concept of 'freedom from want' is used to justify using 'accessibility' and 'availability' as qualitative indicators to assess the impact of political instability leading to migration and refugees.

Section one explores the existing water security literature in order to redefine the term taking in consideration the political instability leading to refugees. The refugee component highlights the urgency of securing basic human needs including access to safe drinking water. In this context, a human-centric approach is found to be the most appropriate and water security is taken to mean 'access to safe and sufficient water for domestic use and sanitation'.

The next section highlights how resource abundance and socio-political factors have influenced the development of water institutions—in order to focus on household methods of accessing water. The chapter then describes the refugee presence and evaluates the impacts on household water access.

Water Perspectives

In 2002, The Food and Agricultural Organisation of the United Nations (FAO) described water security as, "the main goal inspiring the international community".[i]

[1] Water Resource Officer, CBL Division—FAO of the U.N., Viale delle Terme di Caracalla, 00153 Rome, Italy.
Email: amani.alfarra@fao.org
[2] MA Human Rights, University College London, 7A Herbert Road, NW9 6AJ, London, United Kingdom.
Email: islam.abdelgadir@gmail.com

[i] FAO Legal Office. 2002. Law and Sustainable Development since Rio–Legal Trends in Agriculture and Natural Resource Management. Rome: Available. http://www.fao.org/docrep/005/y3872e/y3872e07.htm#bm07.1. Last Accessed: 22.02.2016.

Two years earlier, a declaration of 120 water-ministers at the World Water Forum stated that the 21st century goal of the international community is to "provide water security".[ii] Despite water security's supposed importance, there remains little agreement as to its definition.

Mark Zeitoun et al. claim, "There may be as many interpretations of water security as there are interests in the global water community".[iii] Perspectives originate from concerns over water scarcity to water services. The UN-Water Council highlights the differing "household and community, to local... national... and international settings."[iv] International treaties recognize a human-security component whilst organisations such as the Global Water Partnership spotlight water security for "agricultural and other economic enterprise."[v]

Indeed, "water security is not a conceptual panacea; it does not... boil down all the ways water interacts with livelihoods, economics and environment into two words."[vi] From the hydro-social to the hydro-technical, water security is but one conduit of the study of water. The majority water security literature is written with an eye for policy prescription, with water practitioners debating water security in order to guide policy formation.[vii] In this chapter, water security is used to measure the impacts of refugees on host communities in an unstable political situation.

This section examines the existing perspectives on water security in order to locate this definition, acknowledging that an acceptance of one perspective comes at the expense of others. Perspectives are framed within their respective 'discourses'— defined by Jürgen Link as "an institutionalised way of talking that regulates... action and... exerts power."[viii]

This section will begin by addressing perspectives of water scarcity, framed within traditional security discourses of securitization. Water productivity perspectives will also be considered, with deference to Integrated Water Resource Management (IWRM) discourses. Following this, a critical security studies approach

---

ii World Water Council. 2000. Ministerial Declaration of The Hague on Water Security in the 21st Century, The Hague, Netherlands: Available: http://www.worldwatercouncil.org/fileadmin/world_water_ council/documents/world_water_forum_2/The_Hague_Declaration.pdf. Last Accessed: 03.04.2016.

iii Zeitoun et al. 2013. Chapter 1: Introduction: A Battle for the Ideas of Water Security. In: Bruce Lankford et al. (eds.). Water Security Principles Perspectives and Practices. 711 Third Avenue, New York, NY 10017: Routledge, pp. 3.

iv UN Water. 2013. Water Security & the Global Water Agenda A UN-Water Analytical Brief, United Nations: Available: http://www.unwater.org/downloads/watersecurity_analyticalbrief.pdf. pp. 1. Last accessed: 17.01.2016.

v Global Water Partnership. February 2010. GLOBAL WATER SECURITY, Global Water Partnership 11151 Stockholm Sweden: Available: http://www.gwp.org/Global/Activities/News/GWP_on_ WaterSecurity_Feb_2010.pdf?epslanguage=en. Accessed: 06.12.2015.

vi Mason, N. 2013. Chapter 12: Easy as 1,2,3 Political and Technical considerations for identifying water security indicators. In: Lankford, B.A., Bakker, K., Zeitoun, M. and Conway, D. (Eds.). Water security: Principles, perspectives and practices. Abingdon and New York: Routledge, pp. 200.

vii Zeitoun et al. 2013. opt.cit, pg. 7.

viii Wodak, R. and Meyer, M. 2008. Critical Discourse Analysis: History, Agenda, Theory, and Methodology (Available at: http://miguelangelmartinez.net/IMG/pdf/2008_Wodak_Critical_ Discourse_Analysis_Ch_01.pdf), pp. 25.

will link discourses within the Copenhagen school to the perspectives around human security. This section identifies the household as the appropriate level of analysis.

Defining Water Approaches

Water scarcity

Much of the water security rhetoric manifests directly around perspectives of water scarcity. Water scarcity is a relative concept that requires unpacking.[ix] Jim Winpenny writes, "Scarcity may have its roots in water shortage, but it may also be a social construct."[x] In Arabic culture, water is deemed sacred and its waste or pollution is considered as sin.[xi] Water also has socio-cultural value due to its necessity for securing livelihoods. It has become an issue for national governments to view sovereignty over water resources as integral not only to self-determination, but also self-survival. Nowhere is this clearer than the Jordan River Basin. Multiple hydro-political texts link 20th century Arab-Israeli disputes to regional water claims.[xii]

The MENA region is characterized by physical water scarcity (see Fig. 12.1). Ninety five percentage of areas in the host communities are made up of arid and semi-arid land.[xiii] The population has tripled in the last 50 years, with water consumption following suit.[xiv] Rising water demands has led to the over-exploitation of regional water resources. In 2005, the Millennium Ecosystem Assessment reported that MENA's 15–35% of irrigation withdrawals exceed sustainable supply rates.[xv]

Forced migration has complicated these issues. Some refugee studies suggest that host communities experience net benefits from refugee influxes. For example, Jennifer Alix-Garcia and David Saah's study on refugee inflows in Tanzania argues

---

[ix] Schutter, O. and Pistor, K. 2015. Chapter 1: Introduction: Toward voice and reflexity. pp. 3–49. In: Pistor, K. and Schutter, O. (eds.). Governing Access to Essential Resources. New York: Columbia University Press.

[x] Winpenny, J. 2013. Managing Water Scarcity for Water Security. Rome: Food and Agricultural Organisation. Paragraph W7.

[xi] Dolatyar, M. and Gray, T.S. 2000. Water Politics in the Middle-East: A Context for Conflict or Co-operation. New York: Palgrave Macmillan.

[xii] Key texts include:
(1) Ohlsson, L. 1995. Hydropolitics: Conflicts over Water as a Development Constraint. Volume 3 edn., London & New Jersey: Zed Books, University Press LTD.
(2) Rabil, R.G. 2003. Embattled neighbours: Syria, Israel and Lebanon, USA: Lynne Ryienner Publishers.
(3) Wolf, A.T. 1995. Hydropolitics along the Jordan River: Scarce water and its impact on the Arab-Israeli conflict, Japan: United Nation's University Press.

[xiii] Brauch, H.G., Liotta, P.H., Marquina, A., Rogers, P. and Selim, M.E.-S. (Eds.). 2003. Security and Environment in the Mediterranean: Conceptualising Security and Environmental Conflicts, Volume 3 edn., Germany: Springer. pp. 659. https://www.springer.com/gp/book/9783642624797#reviews.

[xiv] El-Naser, H. 2009. Management of Scarce Water Resources: A Middle Eastern Experience, Witpress, UK pp. 3.

[xv] World Resources Institute. 2005. Millennium Ecosystem Assessment, 2005. Ecosystems and Human Well-being: Biodiversity Synthesis World Resources Institute, Washington, DC pp. 8. Available: https://www.millenniumassessment.org/documents/document.356.aspx.pdf.

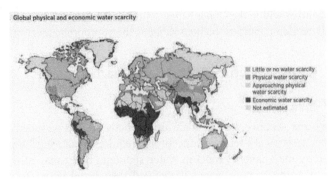

Figure 12.1 World water scarcity.[xvi]

Color version at the end of the book

that wealth benefits resulted from an increased demand for domestic goods.[xvii] This may not be the case in water-scarce contexts. Reports from ALNAP[xviii] and the World Bank[xix] highlight how in host communities with poorer water infrastructure, integration can increase competition over scarce resources.

Likewise, in areas of water scarcity, increased competition can fuel social tensions. Mohamed El Zain relates political conflict in the riparian provinces of Sudan (riparian meaning land situated along a watercourse) as a result of refugee inflows from civil-war regions.[xx]

Water securitisation

Traditional security perspectives relate water to the relationships between state actors in the field of international relations. Through a realist lens, water security becomes synonymous with strategic security. As water becomes necessary for self-survival, or a "national security threat",[xxi] state actors enter a zero-sum game over attaining those resources.

xvi   World Water Development Report 4. March 2014. World Water Assessment Programme (WWAP), As cited in: UN water (2014/11/24) Water Stress versus Water Scarcity, Available at: http://www.un.org/waterforlifedecade/scarcity.shtml (Accessed: 08.04.2016).

xvii   Alix-Garcia, J. and Saah, D. 2009. The effect of refugee inflows on host country populations: evidence from Tanzania. *World Bank Economic Review* 24(1): 148–170.

xviii Ramalingam, B. and Clarke, P.K. 2012. Meeting the Urban Challenge: Adapting Humanitarian Efforts to an Urban World. London: ALNAP/Overseas Development Institute.

xix   World Bank. July 29th 2010. World Development Report 2011 Background Note. The Impacts of Refugees on Neighboring Countries: A Development Challenge, World Bank: Available at: http://siteresources.worldbank.org/EXTWDR2011/Resources/6406082-1283882418764/WDR_Background_Paper_Refugees.pdf (last Accessed: 30.03.2016).

xx   El Zain, M. 2008. Chapter 7: People's Encroachment onto Sudan's Nile Banks and its impacts on Egypt. pp. 129–161. *In*: Nevelina I. Pachova, Libor Jansky, Mikiyasu Nakayama, Shuntaro Yamamoto, Naho Mirumachi, Marko Keskinen, Katri Mehtonen, Olli Varis, Virpi Stucki, Madiodio Niasse, Mahmoud El Zain, Bastien Affeltranger, Kayo Onishi, Ayșegül Kibaroglu, Anthi D. Brouma, Mete Erdem, Jeroen F. Warner, Richard Meissner (eds.). International Water Security. Hong Kong: United Nations University Press.

xxi   Lankford. 2013. Chapter 21: A Chapter: The incodys water security model. pp. 337. *In*: Lankford et al. (eds.). Water Security Principles Perspectives and Practices. 711 Third Avenue, New York, NY 10017: Routledge.

Water becomes the object of securitisation, with "the lack of access to water (being) viewed as a political issue, as the very object of securitisation".[xxii] Water securitisation is a dynamic and controlling endeavour—making water the object of a security concern, and then pursuing its appropriation. For Barry Buzan et al., securitisation is the more extreme version of politicization. The authors envision a spectrum, in which water as a public issue ranges from non-politicization (not a concern of the state) and politicization (a central component of public policy and governance) to securitisation (an existential threat requiring emergency measures).[xxiii]

Water in the Middle East has been historically been securistised. Ismail Serageldin, the former World Bank Vice President for Environmentally Sustainable Development, claimed that the "wars of the next century will be over water."[xxiv] Dr. Boutros Boutros-Ghali, the former UN Secretary-General, agreed, "The next conflict in the Near-East will be about the question of water."[xxv]

With these Neo-Malthusian narratives, Malin Falkenmark and Jan Lundqvist declare that states are prepared to wage war over water supplies.[xxvi] However states in the Middle East have so far not engaged in water conflict. There is a consensus among hydro-political authors that co-operation over water resources is the norm.[xxvii]

It is important to investigate why water has been viewed strategically, and how securitisation discourses have influenced perceptions of water security. One insight is the concept of 'tactical securitisation'—where low-politics issues, such as water sovereignty, are linked with high-politics issues, such as national survival. For example, in the 1994 peace treaty between Israel and Jordan, motivation for establishing a water agreement was vested in national security, or high politics.[xxviii]

In other words, the rhetoric understands achieving water security as a conduit for achieving wider environmental and national security. As discussed in Chapter 1 there are connections between water security and other national concerns in the

---

[xxii] Leb and Wouters. 2013. Chapter 3: The water security paradox and international law: securitisation as an obstacle to achieving water security and the role of law in describing the world's most precious resource. pp. 26. *In*: Bruce Lankford, Karen Bakker, Mark Zeitoun and Declan Conway (eds.). Water Security Principles Perspectives and Practices. 711 Third Avenue, New York, NY 10017: Routledge. http://www.gbv.de/dms/tib-ub-hannover/744175011.pdf.

[xxiii] Buzan, B., Wæver, O. and De Wilde, J. 1998. Security: a new framework for analysis. Lynne Rienner Publishers. pp. 23.

[xxiv] Financial Times, 7 August 1995, as quoted in Mustafa Dolatyar and Tim Gray, Water Politics in the Middle East: A Context for Conflict of Co-operation, London, Macmillan, 2000, pp. 8, 22. http://users.sussex.ac.uk/~js208/govandopposition.pdf.

[xxv] Translated from "le prochain conflit dans la region du Proche Orient portera sur la question de l'eau" (Al-Ahram, Weekly, 19–25 march, 1992). Cited in: Kunigk, E. (1998/1999) policy Transformation and Implementation in the Water Sector in Lebanon: The Role of Politics. Water Issues Study Group School of Oriental and African Studies (SOAS) University of London, Occasional Paper No. 27, pp. 9 [Online]. Available at: https://www.soas.ac.uk/water/publications/papers/file38370.pdf (Accessed: 20.01.2016).

[xxvi] Falkenmark, M. and Lundqvist, J. 1995. Looming water crisis: new approaches are inevitable. pp. 783. *In*: Ohlsson, L. (ed.). Hydropolitics: Conflicts over Water as Development Constraint. London: Zed Books.

[xxvii] See: (1) Lowi, M. 1993. Water and power: the politics of a scarce resource in the jordan river basin (1995a) rivers of conflict, rivers of peace. *Journal of International Affairs* 49: 123–44. (2) Selby, J. 2013. Cooperation, domination and colonisation: The Israeli-Palestinian Joint Water Committee. Water Alternatives 6(1): 1–24.

[xxviii] Stephan Libiszewski. 1995. Occasional Paper No. 13 Water Disputes in the Jordan Basin Region and their Role in the Resolution of the Arab-Israeli Conflict, Environment and Conflicts Project (ENCOP): pp. 78. Available: http://www.mideastweb.org/Mew_water95.pdf.

"water-energy-food nexus"—the concept of a 'nexus' highlighting the intrinsic link between water security and agricultural or economic development.

Virtual and small water

Nevertheless the MENA region remains water scarce. P.B. Anand considers water scarcity "as a symptom of unsustainable use of water resources".[xxix] Yet despite evident over-exploitation, regional solutions to MENA water shortages are not ranked highly on MENA government's political agendas.[xxx]

Tony Allan's seminal work, *The Middle East Water Question: Hydro-politics and the Global Economy*, provides one explanation. Allan purports that:

> "Water shortages are not determining constraints because there is a ready supply of 'virtual water' available. (Virtual) water is contained in the imports of staple foods which are for the moment readily available and massively subsidized on the international market."[xxxi]

Instead of wasting water for manufacturing food or other essential goods, a country can access these goods via trade or international finance.

Allan also makes the crucial distinction between 'small water' (municipal) and 'big water' (industrial). Big water makes up 90 percentage of total water use in the Middle East.[xxxii] Indeed, agriculture alone accounts for some 70 percentage of global water use.[xxxiii]

Municipal water (water withdrawn for direct use by the population) makes up the remaining 10 percentage, which can be adequately covered by regional resources.[xxxiv] Satisfaction of big water needs can be achieved by appealing to the international market. Hence national issues previously concerning water security, now concern economic security. This represents a paradigm shift in the literature, with water security now the concern of water resource management—or water productivity.

Before addressing water productivity, there has emerged a scaling phenomenon. By highlighting the need to only securities 'municipal water', water security scales down, losing legitimacy at the macro-scale. This provides an incentive for rejecting the macro-level approach of traditionalists. Instead it becomes more appropriate

---

[xxix] Anand, P.B. 2007. Scarcity, Entitlements and the Economics of Water in Developing Countries, UK: Edward Elgar Publishing Limited. pp. 23.

[xxx] Kunigk, E. 1998/1999. Policy transformation and implementation in the water sector in Lebanon: the role of politics. Water Issues Study Group School of Oriental and African Studies (SOAS) University of London, Occasional Paper No. 27, pp. 9. Available at:https://www.soas.ac.uk/water/publications/papers/file38370.pdf (Accessed: 20.01.2016).

[xxxi] Allan, J. 1994c. Water: a substitutable resource in the Middle East. Lecture at Durham University, 19th October. As cited *In*: Dolatyar, M. and Gray, T.S. 2000. Water Politics in the Middle-East: A Context for Conflict or Co-operation. New York: Palgrave Macmillan pp. 24/25.

[xxxii] Allan, J. 2001. The Middle East water question: Hydropolitics and the global economy (p. 220). London: I.B. Tauris. ISBN 1-86064-813-4.

[xxxiii] Clement, F. 2013. Chapter 10: From water productivity to water security: a paradigm shift? *In*: Lankford, B., Bakker, K., Zeitoun, M. and Conway, D. (Eds.). 2013. Water Security: Principles, Perspectives, and Practices. 711 Third Avenue, New York, NY 10017: Routledge, pp. 148.

[xxxiv] Ibid.

to consider the water security of micro-actors, namely households, as the level of analysis. One could argue that taking the 'state' would still be sufficient as national security consists of the collective concern for individual needs. However Zeitoun highlights that individual water security can coexist with national water insecurity (e.g., prosperous citizens in water-scarce Kuwait City), as can national water security with individual water insecurity (e.g., rural subsistence farmers in water-rich Zambia).[xxxv]

IWRM and water productivity

Water productivity has evolved as a measure of how efficiently a system converts water units into goods and services.[xxxvi] Due to water's social value, citizens may not view it as a 'scarce commodity'. Colin Chartes and Samyuktha Varma describe water provision as a government obligation.[xxxvii] Subsidized by their governments, consumers are accustomed to treating water as a service supplied to them at zero or low cost. Economics dictate that rising water demands will raise water prices, with state budgets bearing the costs. A water productivity perspective argues that attributing monetary values to water should alleviate fiscal pressures. Water productivity is a central pillar of IWRM. IWRM maintains significant legitimacy in the international community, arguably being the dominant water management paradigm influencing water security.[xxxviii] Attempts at promoting IWRM have been made at the 2001 International Conference on Freshwater in Bonn and the 2002 World Summit on Sustainable Development in Johannesburg.[xxxix]

IWRM's adherence to 'Productivism'[xl] renders water security deterministic. It, "formulates the water problem as one that is determined by the power over nature rather than by the power of one social group over another."[xli] Although geo-climatic conditions and infrastructure arrangements are important in the organisation of water management systems, these physical characteristics cannot be separated from social

---

[xxxv] Zeitoun, M. 2013. Chapter 2: The web of sustainable water security. pp. 11–25. *In*: Lankford, B., Bakker, K., Zeitoun, M. and Conway, D. (Eds.). 2013. Water Security: Principles, Perspectives, and Practices. 711 Third Avenue, New York, NY 10017: Routledge.

[xxxvi] Molden, D. 1997. Accounting for water use and productivity. International Irrigation Management Institute, Colombo. As cited in: Clement (2013) op.cit pp. 149.

[xxxvii] Chartres, C. and Varma, S. 2011. Out of Water: From abundance to scarcity and how to solve the world's water problems, Canada: Pearson Education Limited. pp. 142.

[xxxviii] Cook, C. and Bakker, K. 2011. Water security: Debating an emerging paradigm. *Global Environmental Change* (22): 94–102. Available at: http://web.mit.edu/12.000/www/m2017/pdfs/watsec.pdf. Accessed: 16.04.2016.

[xxxix] Rahaman, M., Varis, O. and Kajander, T. 2004. EU water framework directive vs. integrated water resources management: the seven mismatches. *Water Resources Development* 20(4): 9 [Online]. Available at: http://www.internationalwaterlaw.org/bibliography/articles/general/Rahaman-EUWFD.pdf (Accessed: 16.03.2016).

[xl] 'Productivism' accepts productivity and growth as ultimate goods for organising society. Source: Ozzie Zehner. 2012. Out of Water: From abundance to scarcity and how to solve the world's water problems, The Dirty Secrets of Clean Energy and the Future of Environmentalism edn., Lincoln and London: University of Nebraska Press.

[xli] Swyngedouw, E. 2004. Social Power and the Urbanization of Water: Flows of Power. New York: Oxford University Press. pp. 176.

relations.[xlii] As such, "an economic approach to water is potentially subject to criticism from other social sciences for ignoring many dimensions of agency and focusing on methodological individualism based on a conception of rational choice."[xliii]

Any adopted definition of water security should avoid apolitical approaches. Although IWRM is integrative in terms of science rationality and planning, it defines water security through prescriptive means and simplifies social realities.

Water and human security

Two perspectives of water security have been identified: water scarcity and water productivity. Both provide valuable insight into existing narratives but fail to be an adequate analytical tool for the study of this dissertation. Water securitisation is useful for identifying attendant security concerns but its macro-level focus is inappropriate given the prominent refugee component. Water productivity is useful in examining the narratives around the water provision. However it is fundamentally apolitical, holding more relevance for prescribing policy than describing impacts.

As this chapter does not intend to propose a master plan for water supply, a more critical approach must be taken. The Copenhagen school broadens the security agenda to ask more fundamentally how the idea of 'security' itself is understood. This is enamoured by the question: "water security for whom"[xliv]

David Grey and Claudia Sadoff advocate a definition that captures the more human-centric notions of security.[xlv] In his book *Critical Security Studies and World Politics*, Ken Booth highlights that traditional security perspectives neglect individual human realities.[xlvi] With reference to the human security literature, adopting a 'human-centric perspective' of water security could overcome this. Additionally, this perspective would reject the empiricist approach of water productivity and instead adopt normative approaches towards sustaining livelihoods.

This notion of 'sustaining livelihoods' connects to Amartya Sen's theory of capabilities, which emphasizes empowerment and expanding individual freedoms.[xlvii] In November 2002, The UN Committee on Economic, Social and Cultural rights set out a document describing the human right to water as indispensable for leading a life in human dignity.[xlviii] UN-Water defines water security as "the capacity

[xlii]   Swyngedouw, E. and UNDP. 2006. Power, Water and Money: Exploring the Nexus, Human Development Report Office OCCASIONAL PAPER: UNDP pp. 27.

[xliii]   Anand, P.B. 2007. op.cit, pp. 16.

[xliv]   Harrington and Cameron. 2013. Fluid Identities: Toward a Critical Security of Water. Electronic Thesis and Dissertation Repository. Paper 1716. pp. 84.

[xlv]   Grey, D. and Sadoff, C.W. 2007. Sink or swim? Water security for growth and development. *Water Policy* 9: 545–571.

[xlvi]   Booth, K. (ed.). 2005. Critical Security Studies and World Politics Boulder, CO: Lynne Rienner Publishers.

[xlvii]   Amartya K. Sen. 1999. Development as Freedom. New York: Knopf. As cited in: King, G. and Murray, C.J.L. 2001. Rethinking Human Security. *Political Science Quarterly* 116: 585–610. doi: 10.2307/798222 pg 585.

[xlviii]   UNESCO. 2002. Twenty-ninth session Geneva, 11–29 November 2002 Agenda item 3_Substantive Issues Arising in the Implementation of the International Covenant on Economic, Social and Cultural Rights, General Comment No. 15 pp. 1. Available at: http://www2.ohchr.org/english/issues/water/docs/CESCR_GC_15.pdf. Last accessed: (29.11.2015).

of a population to safeguard sustainable access to adequate quantities of acceptable quality water for sustaining livelihoods."[xlix] Within the human security literature, these examples relate to the notion of securing 'freedom from want'.[l]

The refugee component of this study highlights the urgency of securing basic human needs. Indeed, the first premise of the ministerial declaration on Water Security at the Second World Water Forum in 2000 was "meeting basic human needs."[li] 'Freedom from want' sees water security as the capacity to realize those needs.

For many organisations, including the World Health Organisation and UN-Habitat, this assumes water for domestic use and sanitation[lii]—widely considered as minimum water requirements. As such, water security can be taken to mean 'access to sufficient and safe water for domestic use and sanitation'.

This definition receives legitimacy from the international community. In 1977, the UN Water Conference proclaimed the 1980's as the International Water Supply and Sanitation Decade, this was approved unanimously in 1980 by the UN general assembly.[liii] The human right to water itself was explicitly acknowledged in 2010 through UN Resolution 64/292.[liv]

A topical criticism of the human-centric approach is its identification with "western values".[lv] Given that this dissertation is partly motivated by a rejection of Eurocentric positions, perhaps a human-centric analysis of both Jordanian and Lebanese contexts should be cautioned. On the other hand, Philip Jan-Schafer remarks that numerous high-ranking Jordanian officials have endorsed 'freedom from want'.[lvi] Karim Makdisi also notes that in Lebanon this approach is "politically accepted" and "customarily guaranteed".[lvii]

Human security critics highlight it as vague and difficult to measure. Roland Paris argues that the concept is "slippery by design".[lviii] Nicholas Thomas and

---

[xlix] UN Water. 2013. opt.cit, pp. vi.

[l] NEWMAN, EDWARD. 2010. Critical human security studies. Review of International Studies Cambridge University Press. 36(1): 77–94. http://www.jstor.org/stable/40588105.

[li] HRH The Prince of Orange, Frank R. Rijsberman. 2000. Summary report of the 2nd World Water Forum: from vision to action. Elsevier Science Ltd: Water Policy 2: 387–395. Available at: http://www.yemenwater.org/wp-content/uploads/2013/03/HRH-The-Prince-of-Orange-and-Frank-R.-Rijsberman.-2000..pdf. Last accessed (04.04.2016).

[lii] OHCHR. 2003. The right to water, Fact Sheet No. 35: pp. 3. Available at: http://www.ohchr.org/Documents/Publications/FactSheet35en.pdf. Last Accessed (12.04.2016).

[liii] Brooks, D. 2008. Human rights to water in North Africa and the Middle East: what is new and what is not; what is important and what is not. pp. 24. *In*: Biswas, A., Rached, E. and Cecilia Tortajada (eds.). Water as a Human Right in the Middle East and North Africa. New York: Routledge Taylor and Francis Group.

[liv] Ibid pp. 26.

[lv] Tadjbakhsh, S. and Chenoy, A. 2007. Human Security: Concepts and Implications. London and New York: Routledge Taylor & Francis Group. pp. 158.

[lvi] Schäfer, P.J. 2013. Chapter 3: Water security. *In*: Philip Jan Schafer (ed.). Human and Water Security in Israel and Jordan. Weisenheim am Berg Germany: Springer, pp. 49.

[lvii] Makdisi, K. 2008. Towards a human rights approach to water in Lebanon: implementation beyond "reform". pp. 166. *In*: Biswas, A., Rached, E. and Cecilia Tortajada (ed.). Water as a Human Right in the Middle East and North Africa. New York: Routledge Taylor and Francis Group.

[lviii] Roland Paris. 2001. Human security: paradigm shift or hot air? International Security 26(2): 97. As cited in Booth, K. (2007). Theory of World Security (Vol. 105) Cambridge University Press pp. 157.

William Tow advocate that narrowing the definition of human security would increase its analytical capacity.[lix] In their reply, Alex Bellamy and Matt Mcdonald caution that this exercise would betray human security's expansive foundations.[lx] Ultimately, a human-centric approach to water security widens the parochial focus of traditionalists whilst narrowing the lens of human security. Resulting in an apt trade-off between parsimony and explanatory power.

All things considered, this literature review finds "access to sufficient and safe water for domestic use and sanitation" as an appropriate definition for this dissertation. The human-centric approach takes into consideration existing social and political nuances in a way that is consistent with a refugee context.

Household approaches

In *Security: A New Framework for Analysis*, Buzan et al. spotlight the use of 'levels of analyses in order to locate the actors, referent objects and dynamics of interaction that operate in the realm of security.'[lxi] The 'referent object' denotes what is considered as existentially threatened. A traditional security perspective considers the state as the referent object. As discussed previously, this perspective is marginalised by way of Allan's concept of 'small water'. Instead, the household has been taken as the referent object. The household approach adopted here inverts the level of analysis, considering the household as the referent object and the state instead as the securitising actor. As such, the household becomes the source of explanation for understanding water security outcomes.

Given the scale of the refugee influx and the unlikelihood of repatriation, refugee households can be included in the analysis as 'part of the host community'. One limitation of the household approach is the neglect shown to those with insecure shelter, which is of particular concern regarding refugees. Cursory attention will be paid towards the water security of those in 'informal settlements' (such as collective shelters, unfurnished houses or improvised lodging).[lxii] Unfortunately, this attention is limited due to the lack of available information.

Relating water security to 'freedom from want' denotes a positive conception of water security. For David Brooks, "water rights mean entitlements—something that one gets (or should get) by legal authority."[lxiii] Although the human-centric approach

---

[lix]   Thomas, Nicholas and William T. Tow. 2002. The utility of human security: sovereignty and humanitarian intervention. *Security Dialogue* 33: 177–192.

[lx]   Bellamy, Alex and Matt McDonald. 2002. The utility of human security: which humans? what security? a reply to thomas and tow. *Security Dialogue* 33: 373–377.

[lxi]   Buzan, B., Wæver, O. and De Wilde, J. 1998. op.cit.

[lxii]   UNHCR. February 2014. Inter-Agency Shelter Sector Working Group-Lebanon Shelter Strategy for 2014, Download available at: https://www.google.co.uk/https://www.google.co.uk/url?sa= t&rct=j&q=&esrc=s&source=web&cd=1&cad=rja&uact=8&ved=0 ahUKEwiU0tWa9bjMAhWJbRQKHfmyCSkQFgggMAA&url=https percentage 3A percentage 2F percentage2Fdata.unhcr.orgpercentage2Fsyrianrefugeespercentage2Fdownload.phppercentage3Fid percentage 3D4582&usg=AFQjCNFbTM2RltMl9BFjUgHx_3TRSwIpgA&sig2=3yAa7wA82n_ UHs6dwXrIiA (Accessed: 23.04.2016).

[lxiii] Brooks, D. 2008. op.cit, pp. 20.

rejects a state-centric analysis, Nick Vaughan-Williams highlights the pertinent role played by the state in the provision of water security.[lxiv] Moreover, the Centre for Economic and Social Rights clarifies that states have to "promote, facilitate and provide individuals with access to adequate water in terms of availability, quality and accessibility."[lxv]

Buzan et al. add that casual explanations of security outcomes result from linking relevant levels of analysis to each other. For example, top-down (from system structure to unit-behaviour) or bottom-up (from human nature to the behaviour of human collectives).[lxvi] The former identifies a critical institutional analysis, whilst the latter denotes a rational-choice approach. As adequate water provision can be considered a state obligation, this dissertation utilizes the former approach as its research focus, mapping the dynamics between water institutions and household water security.

Conclusion

The above discussion established water security as 'access to safe and sufficient water for domestic needs and sanitation'. Adopting a human-centric perspective, the research methodology took the household as the level of analysis. Relating water security to 'freedom from want', the qualitative indicators of *accessibility* and *availability* were identified in order to guide the assessment.

Adopting a specific definition or perspective, or deciding on a set of indicators, necessitated an omission of other methodologies. Other areas for research include how refugees have affected the region's hydro-politics, or how changing water demands influence food and energy sectors. Notably, the refugee component emphasizes the urgency of securing basic human needs and the adopted definition reflected this. This is further supported by the acceptance of humanitarian concerns within international water security discourses.

The indicators of accessibility and availability guided a critical institutional analysis considering the political economies surrounding water provision and how they influence access. Many socio-political realities were alluded to in this process, often briefly.

If recommendations are to be made based on this study, they should be directed towards two groups: the international community and water policy practitioners. While efforts are focused on tackling the crisis north of the Mediterranean,[lxvii] only 43.1 percentage of the UN's 2015 Syria Response Plan appeal was funded.[lxviii] In dealing with the crisis, the international policy has chosen traditional security

---

[lxiv]   Peoples, C. and Vaughan-Williams, N. 2014. Critical Security Studies: An Introduction. Routledge.

[lxv]   UNESCO. 2002. op. cit, pp. 18. paragraph 59.

[lxvi]   Buzan, B., Wæver, O. and De Wilde, J. 1998. op.cit.

[lxvii]   Council of the European Union (18/03/2016 Press Release Foreign affairs & international relations) EU-Turkey statement, 18 March 2016, www.mercycorps.org: Available at: http://www.consilium. europa.eu/en/press/press-releases/2016/03/18-eu-turkey-statement/# (Last Accessed: 12.04.2016).

[lxviii]   Financial Tracking Service. 2015. Syria Response Plan 2015 (HRP) snapshot 2015, Available at: https://ftsbeta.unocha.org/appeals/1069/summary (Accessed: 17.04.2016).

discourses over human security discourses. The study of water security provides a fresh lens into the neglected outcomes. Indeed, as water scarcity becomes more of a regional concern, water policy practitioners should consider the impacts on basic household needs. In that sense, the SRC has also provided a useful avenue through which to consolidate water security's humanitarian responsibilities towards sustaining livelihoods.

Taking into consideration the water security for refugees and the needs to provide them with minimum needs in water, usually we forget the negative impact of sanitation and the negative impact on water quality.[lxix]

[lxix]   UNHCR. 2005. Water Manual for refugee situations. http://ec.europa.eu/echo/files/evaluation/watsan2005/annex_files/UNHCR/UNHCR5%20-%20Water%20Supply%20in%20Refugee%20Situations.pdf.

Chapter-13

In the Context of Crisis and Change
Governing the Water-Energy-Food Nexus in the Jordan River Basin

Amani Alfarra,[a,*] *Islam Abdelgadir*[b] and *Velma I. Grover*

INTRODUCTION

The Middle East[1] is one of the most water scarce regions in the world and pressure on water resources is likely to increase with exploding populations, rapid urbanization, expansion of the agricultural sector and soaring demands of a more affluent society.

Water scarcity is increasingly affecting the economic and social development of the region's countries, the region has 6.3 percent of the world's population but only 1.4 percent of the world's renewable fresh water (Roudi-Fahimi et al., 2002). Further, many of the region's water resources, surface and groundwater, are shared among riparian states in the different watersheds in the region. A strong and good water management approach is needed to solve the problems related to both water quality and quantity.

Irrigated agriculture is the main source of water consumption, accounting for 85 percent of water withdrawals in the region (Al-Zubari, 2015). The overuse of water in agriculture negatively affects the countries' already limited water resources. Moreover, Food and Agriculture Organization (FAO) has anticipated that the global expansion of irrigation would be strongest (in absolute terms) in the land-scarce

[a] Water Resource Officer, CBL Division—FAO of the U.N., Viale delle Terme di Caracalla, 00153 Rome, Italy.
[b] MA Human Rights, University College London, 7A Herbert Road, NW9 6AJ, London, United Kingdom. Email: islam.abdelgadir@gmail.com
* Corresponding author: amani.alfarra@fao.org
[1] As defined here, the Middle East and North Africa region includes Algeria, Bahrain, Egypt, Iran, Iraq, Israel, Jordan, Kuwait, Lebanon, Libya, Morocco, Oman, Palestine, Qatar, Saudi Arabia, Syria, Tunisia, Turkey, the United Arab Emirates, and Yemen.

regions that are hard-pressed to raise crop production by intensive cultivation practices. In the Near East and North Africa this is expected to increase by an additional 6 million ha. Meanwhile, further expansion will become increasingly difficult as water scarcity increases and competition for water from other stakeholders: households and industry continue to reduce the share available to agriculture.

The water issues in the region are intrinsically linked to other significant development sectors, such as human health, food, energy, and other interdependencies that carry within them cross-cutting issues of human rights, social, economic, legal, technical, political and security (Al-Zubari, 2014). However, energy, land management and water resources planning commonly take place in isolation, without adequate consideration of what are the planned developments in other sectors, and implications of the development—positive or negative. Lack of inter-sectorial coordination is a major challenge both on the national and inter-basin levels, in developing as well as in developed countries. In inter basin watershed setting, the trade-offs and externalities may cause friction between the riparian countries and different interests. A nexus (or inter-sectorial) approach to managing the interlinked resources can enhance water, energy and food security by increasing efficiency, reducing trade-offs, building synergies and improving governance across sectors. The question is: Which policies, technologies, and actions can help to move in that direction? In Jordan; despite the neutral or even positive impact on labour markets and growth, a trade-off between these impacts and the added strain on public finances will arise in the short-run (Fakih and Marrouch, 2015). The Syrian conflict is severely affecting economic, social, environmental and human development in neighbouring countries, uprooting millions of Syrians and eroding the very foundations of their livelihoods. The security situation of people in Syria and neighbouring Lebanon, Jordan, Iraq, Turkey and Egypt is critical and worsening daily.

Syrian Crisis: Current Situation and its Impact on Neighbours

Currently, in Syria, food supplies are limited, difficult to access and costly. Fields and farming assets have been left idle or destroyed due to violence, displacement, increased production costs and a lack of basic farming supplies. For instance, in 2013, the wheat harvest fell 40 percent short of an average year (FAO, 2013). Further, the Syrian pastoral lands and natural resources are degrading, food chains are disintegrating—from production to markets—and entire livelihood systems are collapsing. Essential services are on the brink of collapse under the burden of continuous assault on critical water and energy infrastructure (Vidal, 2014). The Syrian crisis has had repercussions on economic conditions for countries along the Jordan River Basin, as agricultural production and associated trade activities have been disrupted by the conflict.

The recent, influxes of migrants, particularly to Jordan and Lebanon, have put increasing pressures on the built and natural environment and require a rethinking of the planning, allocation and uses of natural resources. With 6.5 million Syrians internally displaced and a further 2.6 million are registered refugees in Lebanon, Turkey, Jordan, Iraq and Egypt (FAO, Subregional Strategy and Action Plan: Resilient Livelihoods for Agriculture and Food and Nutrition Security in Areas

Affected by the Syria Crisis n.d.), hosting communities face intense competition for resources such as land, water and job opportunities, while costs for housing, energy, food and other commodities soar. The increasing demand from the bourgeoning Syrian refugee population has significantly impacted the water availability and its frequency. Almost everyone in the Jordan River Basin (JRB) has access to drinking water, but the quantity supplied per capita has recently decreased. In Jordan, for instance, water supply has been reduced to 30 l/p/d (litre/person/day), as compared to the standard of Water Authority of Jordan: 100 l/p/d. This is complicated by the fact that in Jordan 40 to 50 percent of water is still lost through network breakdowns, leakages and illegal tapping (MOPIC, 2014). There are also concerns regarding the potential pollution of aquifers due to increased quantities of unregulated wastewater discharge and over pumping.

While not, the primary driver for the regional energy challenges, the current Syrian crisis has exacerbated the situation due to rising local demand and energy costs. In Jordan, the Syrian refugee influx has significantly exacerbated levels of residential energy consumption. For instance, the Syrian refugees have led to an increase of Jordan's residential energy consumption from 5.9 percent in 2011 to of 9.44 percent in 2011 with an associated subsidy cost of Jordanian Dinar (JD) 44.3 million (MOPIC, Needs Assessment Review of the Impact of the Syrian Crisis on Jordan, 2013).

According to the United Nation's High Commissioner for Refugees (UNHCR), the Syrian Refugee Crisis (SRC) represents the worst humanitarian crisis of the post-world war era (Edwards and Dobbs, 2014). Since the civil war began in 2011, it has remained the crisis du jour and the topic of significant discussion and analysis. Unfortunately, much of the focus remains on the impact on European countries. This eurocentric focus distracts from the extent of the impact in the region itself. The UNHCR confirms that the countries with the largest number of Syrian refugees are, Turkey, Jordan and Lebanon. These three countries alone have received nearly 4.3 million Syrian refugees (3RP, 2016); quadruple the number of applications in Europe (UNHCR, 2016).

It is in this context of global change and crisis, there needs to be focus on the impacts on water, energy and food security in the riparian countries of the Jordan River Basin, namely Jordan, Lebanon, Palestine and Syria. Without taking into account the interconnections among the sectors, resource allocations may easily be seen as a zero-sum game where intense competition for resource access can easily lead to a conflict. This chapter seeks to explicitly address the interlinkages of the water sector with other development challenges in the region, in an effort to promote effective cooperation within the basin. This involves the harmonization of national water, energy and food policies, the development cooperation initiatives and the support of investment in the implementation of joint protection and management plans.

Jordan and Lebanon are taken as case studies for this assessment. The two countries have received similar numbers of refugees (UNHCR estimates 638,633 in Jordan and 1,055,984 in Lebanon) and are both considered upper-middle income countries (WB, 2015). However, they have differing water inheritances. Lebanon is objectively water secure—with a water allocation that is feasibly capable of supplying domestic demands (Amery, 2000). Jordan, in comparison, is ranked the fourth most water scarce country globally (UNESCO, 2014). Lebanon is a republic with heavy ethno-religious diversity, while the Jordanian monarchy rules over a relatively homogenous population. The two are comparably influenced by the crisis but are sufficiently distinctive that a comparison should provide a refreshing insight.

A lot of studies have focused on the experiences and livelihoods of refugees themselves. Whilst these contributions are valuable, it is also important to consider impacts on the host communities. With this in mind, the analysis in this chapter is done at the 'household' level. The justifications for a household focus are elaborated in the literature review section; however, one advantage is its ability to place water security within a wider institutional framework of government provision. The research methodology takes a critical institutional approach in examining this water provision. 'Critical Institutionalism' explores how institutions dynamically mediate relationships between people, natural resources and society (Cleaver and Koning, 2015). This approach provides a robust understanding of how forced migration has affected household water security by taking the SRC as an exogenous shock to the existing framework of water institutions.

The chapter argues that the SRC has negatively affected water security in both Jordan and Lebanon, with the impact in Lebanon being more acute due to the institutional setup. Other chapters in this section have defined water security, this chapter will re-define the existing water security literature in order to find the most useful definition for this argument. The refugee component highlights the urgency of securing basic human needs. In this context, a human-centric approach is found to be the most appropriate and water security is taken to mean 'access to safe and sufficient water for domestic use and sanitation'.

The next section will build a context for water provision in Jordan and Lebanon—considering how resource abundance and socio-political factors have influenced the development of water institutions—in order to highlight household methods of accessing water. This is in order to be able to describe the refugee presence/ evaluate the impacts on household water access. Finally, the authors will compare each country in order to understand how the SRC has impacted water security, with analysis given to how water institutions have influenced the nature of this impact.

Major Challenges in the JRB

Geopolitics

The question of water sharing in the Jordan River Basin is inextricably linked to the ongoing regional conflicts and instability, and while a wide range of issues are at stake, control over water in the basin has added to existing regional tensions.

The instability in the region creates various stresses on the natural resources. For instance, forced migration becomes the main stress within the Jordan River Basin. Based on UNHCR figures (2014), Jordan is hosting 1,438,440 refugees.[2] The chapter looks at the types of impact the influx of Syrian refugees will have on the water

---

[2]   http://www.unhcr.org/528a0a2c13.pdf.

sector in the region. Groundwater depletion was found to be a major concern for Jordan's water resources prior to the refugee influx, since the total water extraction rates exceeded the renewable water amount.

International response to Syria crisis

As the crisis shows no signs of abating, the response is now shifting from humanitarian relief to a longer-term, development-focused response. In Jordan the shift has taken place in the form of the 2015 Jordan Response Plan for the Syria Crisis[3] and the 2014–2016 National Resilience Plan.[4] These plans encompass a number of sectors in an attempt to mitigate the impacts of the crisis as well as to build back more resilient livelihood systems. They mark a shift in the way we deal with resources. While basic access to water, energy and food supply still has priority, the focus is now also on the development of sound water supply and establishing water management structures. Nexus thinking can help identify synergies, for example, in the Zaatari Camp in Jordan, where basic water and energy resources are needed on a continuing basis by the 84,000 people currently living in the camp.[5] Renewable energy capacities are being developed to power two planned wells to access the groundwater resources on which the camp is situated. As the crisis is seemingly protracted, we are at the cross roads where humanitarian and development communities should work together more constructively to ensure that their programmes are reinforcing and not conflicting or disrupting each other.

To what extent do national development strategies need to be adjusted to deal with the extra strains the Syrian crisis has put on resources, and more importantly, how can the needs of the increasing number of refugees be met in the long-term?

Water quantity and quality

Ensuring adequate quantities of water for riparian's is a key challenge in the basin given the relatively small volume of water available and the large population. River flow has been greatly reduced over the years as a result of increased exploitation of water resources in the basin. Water quality rapidly deteriorates along the course of the Jordan River and its lower portion is extremely high in salinity and heavily polluted. The quality of water in the Jordan River has severely deteriorated in recent decades, especially, because of the rapid immigration within the basin that has caused stress on the resources mainly in Jordan in light of Jordan's population increase by more than a quarter due to the influx of Syrian refugees.

Energy security

Rapid population growth, rising standards of living and growing energy needs associated with increasing water scarcity have caused regional electricity demand to

---

[3]   http://reliefweb.int/sites/reliefweb.int/files/resources/JRP16_18_Document-final+draft.pdf.
[4]   http://jordanembassyus.org/news/final-draft-national-resilience-plan.
[5]   http://www.academia.edu/14929981/Can_saving_lives_save_livelihoods_The_water-_energy-food_ nexus_and_human_security.

increase at very high rates. For example, electricity consumption rose by 6 percentage in Syria and Lebanon from 2007 to 2008, and by more than 8.5 percentage in Jordan over the same period of time.[6] In JRB, the fossil fuels including petroleum products, natural gas, and coal, are currently the main sources of electricity that are imported from other countries. In addition to creating significant challenges for regional efforts to address global climate change, this reliance on imported fossil fuels also imposes high economic costs and undermines energy security throughout the Jordan River Basin. Despite the region's current reliance on fossil fuels, the Jordan River Basin has a vast renewable energy potential, mainly in the form of direct solar energy that is only just beginning to be tapped.

Food security

The MENA region is the most food import–dependent region in the world, and net food imports are projected to rise even further in the future. Net food imports have accounted for 25–50 percent of national consumption, and this high reliance on imported food can be attributed to both demand- and supply-side factors.[7] Demand-side factors include rising population and changing consumption patterns due to higher disposable income. Shifting demand patterns from staples to higher-value food products, combined with limited potential for land expansion, will further increase the region's food trade deficit. Supply-side factors include limited natural resources such as land and water.

Overall, the Jordan River Basin has an estimated total irrigated area of 100,000–150,000 ha of which around 30 percent is located in Israel, Jordan and Syria, 5 percent in Palestine and 2 percent in Lebanon.[8]

Climate change

The predicted impacts of global climate change could very well be the final impetus needed to push the Jordan River Basin to engage in cooperation on water and energy issues. Current climate models suggest that global climate change will cause the world's most water scarce region to become even hotter and drier. According to the United Nation's Intergovernmental Panel on Climate Change, average temperatures in the Jordan River Basin are expected to increase by up to 3.1 degrees Celsius in the winter and as much as 3.7 degrees Celsius in the summer, which would cause average rainfall to decrease by 20–30 percent over the next thirty years.[9] Climate scientists predict that this would result in further reductions to the Jordan River's flow, a decrease the recharge rates of natural aquifers, cause desertification of arable land and soil erosion along the Mediterranean coast, and increase unpredictability of climatic events.

---

[6]   http://www.geni.org/globalenergy/research/water-energy-nexus-in-the-jordan-river-basin/the-jordan-river-basin-final-report.pdf.

[7]   http://www.fao.org/nr/water/docs/WSRE_Preliminary-Regional-Review-Gap-Analysis-Draft-Report.pdf.

[8]   https://waterinventory.org/sites/waterinventory.org/files/chapters/chapter-06-jordan-river-basin-web.pdf.

[9]   http://www.geni.org/globalenergy/research/water-energy-nexus-in-the-jordan-river-basin/the-jordan-river-basin-final-report.pdf.

Defining water availability

Water availability is defined by the supply of a sufficient amount of water of a potable quality. Sufficient is recognized as a subjective verb, as what is sufficient for one household may be deemed insufficient for another. For the purposes of simplicity in this chapter authors will consider the World Health Organisation (WHO) recommendation of between 50–100 litres per person per day.[10] Analysis will be done on how refugees affect the level of water available and the quality of that water.

Defining water accessibility

Water accessibility is defined by how exactly households receive water from water providers. Consideration will be placed on the commodification of water (the extent to which one's access is related to the cash-nexus). To calculate this, the chapter will compare the level of private water consumption to that of public water. The degree to which governments subsidize water will also be considered. Analysis will be done on how refugees affect the functioning of those water providers and household's relationship to the cash-nexus.

Jordan

Water resources

In a global context, Jordan is characterized as a "water scarce" country. Internationally a "water scarce country" is one with less than 1000 cubic metres of fresh water per person per year (FAO, Water Resources of the Near East Region: A Review 1997) (Winpenny, 2000). Jordan is ranked the fourth most water scarce country in the world.[11] Comprising mostly of desert land with an annual per capita share of 140 cubic metres in 2012.[12] According to United Nation's Educational, Scientific and Cultural Organisation (UNESCO), this constitutes "absolute water scarcity", sitting well below the *water poverty line* of 500 cubic metres.[13] Water supply is sourced from surface and groundwater sources, rainwater collection and desalination plants. The region includes 15 surface water sources and 12 groundwater basins sharing inter-related aquifers.[14] Fluctuating precipitation[15] and international watercourses contribute to surface water.

---

[10] http://apps.who.int/iris/bitstream/10665/44102/1/9789241597906_eng.pdf.

[11] UNESCO Office in Amman. 2014. loc.cit.

[12] Source: MWI. 2009. Water for life: Jordan's water strategy 2008–2022, Water Management pp. 8. Available for download at: http://inform.gov.jo/en-us/By-Date/Report-Details/ArticleId/64/smid/420/ArticleCategory/217/2008-2022-National-Water-Strategy Last Accessed: (12.02.2016).

[13] UNESCO. 2012. Managing Water under Uncertainty and Risk, The United Nations World Water Development Report 4 Volume 1 pp. 140: Available for at: http://www.unesco.org/new/fileadmin/MULTIMEDIA/HQ/SC/pdf/WWDR4%20Volume%201-Managing%20Water%20under%20Uncertainty%20and%20Risk.pdf. Accessed: (28.02.2016).

[14] An aquifer is an underground layer of water bearable rock, from which water wells can extract groundwater. Source: Ibid pp. 11.

[15] MWI. March 2014. Establishing the Post-2015 Development Agenda: Sustainable Development Goals (SDG) towards Water Security The Jordanian Perspective, The Hashemite Kingdom of Jordan pp. 4. Available at: http://www.mwi.gov.jo/sites/en-us/Hot%20Issues/SDG_Jordan%20Precspective_Post%202015.pdf. Last Accessed: (20.03.2016).

In 2008, agriculture constituted over 60 percent of total water demand,[16] and municipal water accounted for 32 percent. The municipal sector is characterized by a sustained chronic water deficit, with some sections of the network losing over 50 percent of water to leakages and illegal connections.[17]

The water situation in Jordan is complex and unsustainable. Jordan experiences growing freshwater demands that already exceed availability from combined surface and groundwater resources (Schyns et al., 2015). Jordan's water conventional, or natural, water resources originate from rainfall, ground waters, and surface waters.

Annual rainfall for the country at-large is estimated at 266 millimeters (mm) per year but this does vary within different regions of the country. Some areas in the northwest area of Jordan can receive as much as 600 mm of rain per year, while sections of the southern and eastern desert areas receive only around 50 mm per year. It is further suggested that 92 percentage of the rainfall evaporates into the atmosphere (JVA, 2011), leaving the country with an even more meager share of rainfall. The rainfall typically falls only between the months of November and April, with no rainfall for the rest of the months. Rainfall is the only source of recharge for the country's groundwater aquifers so its potential decline in the coming years could pose a significant problem for Jordan (Nortcliff et al., 2008).

As for Jordan's groundwater resources, they account for roughly 54 percentage of the total water supply in the country. There are 12 major water basins as seen in Fig. 13.1, along with each basin's annual safe yield in MCM. The total safe yield of all of the basins is estimated at 275 MCM per year, whereas yearly abstraction can be upwards of 473 MCM (Altz-Stamm, 2012).

As surface water is not considered adequate for drinking purposes,[18] drinking water is entirely sourced from groundwater sites. Current exploitation of these groundwater resources is at maximum capacity and in many cases well above the recognised 'safe yield'.[19] Jordan's 2008–2022 Water Strategy predicts a yearly decrease of aquifer levels by 1–2 m.[20] Zeitoun et al. predict that Jordan will not be able to support its own water needs by 2030 (Zeitoun et al., 2011).

---

[16] Raddad, K. 2005. Water supply and water use statistics in Jordan, Department of Statistics Jordan. pp. 4. Available at: http://unstats.un.org/unsd/environment/envpdf/pap_wasess4a3jordan.pdf. Last Accessed: (30.03.2016).

[17] MWI. 2009. op.cit, pp. 41.

[18] Karen Assaf, Bayoumi Attia, Ali Darwish, Batir Wardam and Simone Klawitter. 2004. Water as a human right: The understanding of water in the Arab countries of the Middle East—A four country analysis Global Issue Papers No. 11 Heinrich Böll Foundation pp. 118. Available at: http://www.emwis.org/countries/fol749974/country769281/PDF/waterbook. Last Accessed: (15.04.2016).

[19] Holst-Warhaft, G. and Steenhuis, T.S. 2010. Losing Paradise: the water crisis in the Mediterranean. Ashgate Publishing, Ltd. pp. 153.

[20] Ministry of Water and Irrigation. January 2015. Jordan Water Sector Facts and Figures 2013, pp. 29. Available at: http://www.mwi.gov.jo/sites/en-us/Documents/W.%20in%20Fig.E%20FINAL%20E.pdf. (Last Accessed: 20.03.2016).

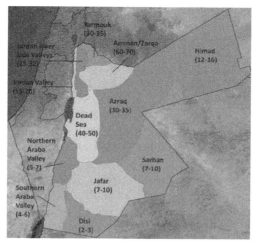

Figure 13.1 Groundwater Basins in Jordan and their annual safe yield in million cubic meters.

Color version at the end of the book

Water governance

The Ministry of Water and Irrigation (MWI), established in 1988, is the highest water authority in Jordan.[21] The Jordan's Ministry of Water and Irrigation is responsible about all water resources supply, management and infrastructure.[22] Management of water provision is delegated to the Water Authority of Jordan (WAJ), an autonomous corporate body holding financial and administrative independence.[23] WAJ owns several water companies (utilities) responsible for water and wastewater related services in the 12 governorates. The four northern governorates with the highest water consumption; Irbid, Ajloun, Jerash and Mafraq, are all supplied by the Yarmouk Water Company.[24]

The GoJ has sought to be proactive on the issue of water scarcity. The *Water Strategy* states that Jordan "will no longer entertain any irresponsible drain of our water resources".[25] A significant proportion of Jordan's water sector receives funding from international organisations. For example, the World Bank has provided loans for all major water infrastructure projects.[26] As a result of international intervention, water productivity and Integrated Water Resources Management (IWRM) perspectives have influenced water policy.

[21] Beaumont, P. 2005. Water Institutions in the Middle East. pp. 147. *In*: Biswas, A., Tortajada, C. and Gopalakrishnan, C. (eds.). Water Institutions: Policies, Performance and Prospects. New York: Springer.

[22] Dante Augusto Caponera. 2003. National and international water law and administration: selected writings (Vol. 9). Kluwer Law International. pp. 115.

[23] MWI. 2009. op.cit. pp. 84.

[24] Laurence Hamai, Ruth McCormack, Eva Niederberger and Lou Lasap. March 2013. Integrated Assessment of Syrian Refugees in Host Communities Emergency Food Security and Livelihoods; Water, Sanitation and Hygiene; Protection, OXFAM GB, Jordan: OXFAM pp. 9.

[25] MWI. 2009. op.cit. pp. 32.

[26] Haddadin, M.J. 2006. Water resources in Jordan: evolving policies for development, the environment, and conflict resolution. pp. 254. Resources for the Future.

Water provision and access

In Jordan, 97 percentage of the population is linked to the public water supply. Nearly 100 percentage of urban and 87 percentage of rural populations are served with piped water.[27] Piped water is subject to a progressive tariff that ranges regionally.[28] A blanket cost is charged for the first 18 m³ of water, after which a higher rate applies.[29] Water tariffs in rented accommodation are paid through the landlord. Oxfam finds that renters are charged an average 10–12 JD/month more than homeowners.[30]

Water supplies to households have been rationed since 1987.[31] Water is supplied to neighbourhoods on a rotational basis, mostly once per week during the winter months and at a lower frequency, once a fortnight, during summer.[32] Government subsidization allows piped water to be largely affordable. Although, households have access to drinking water, yet the intermittency means that piped water is inadequate for covering household demands. Many households store water until the next available supply, with average sizes for water containers ranging from 2–6 m³.[33] Oxfam highlights the relationship between water storage capacity and household wealth, with wealthier households having larger water storage facilities or being able to invest in underground tanks.[34]

Jordan's underground water By Law 85 of 2002—emphasizes that selling or distributing water is prohibited without government approval.[35] There are approximately 2000–3000 groundwater wells in Jordan,[36] and well licenses must be yearly renewed.[37] These wells supply private water markets, with private tankers selling and distributing any water extracted. Despite strict water laws, an estimated half of these wells operate illegally.[38] The *wasta* culture incentivizes water officials to pursue personal interests, distributing licenses to familial and tribal ties.[39]

Often water stored does not cover household water needs and many households must supplement those needs through bottled water and privately sold tank water.

---

[27] Karen Assaf, Bayoumi Attia, Ali Darwish, Batir Wardam and Simone Klawitter. 2004. Water as a human right: The understanding of water in the Arab countries of the Middle East—A four country analysis Global Issue Papers No. 11 Heinrich Böll Foundation pp. 118. Available at: http://www.emwis.org/countries/fol749974/country769281/PDF/waterbook. Last Accessed: (15.04.2016).

[28] Brooks, D.H. 2013. Infrastructure Regulation: what works, why and how do we know? Lessons from Asia and beyond. Edited by Darryl, S.L. Jarvis, M. Ramesh, Xun Wu and Eduardo Araral, Jr. World Scientific, Singapore pp. 529.

[29] Wildman, T. 2013. op.cit. pp. 23.

[30] Ibid. pg. 18.

[31] Darmame, K. and Potter, R.B. 2009. Op.cit. pp. 2.

[32] Wildman, T. 2013. Water Market System in Balqa, Zarqa, & Informal Settlements of Amman & the Jordan Valley–Jordan August–September 2013, OXFAM pp. 5.

[33] Laurence Hamai, Ruth McCormack, Eva Niederberger, Lou Lasap. March 2013. op.cit. pp. 11.

[34] Wildman, T. 2013. op.cit. pp. 24.

[35] Government of Jordan. 1988. Law No. 18 of 1988 Water Authority Law, FAO Legal office: Available online at: http://faolex.fao.org/docs/pdf/jor1338E.pdf (Last accessed: 19.04.2016). Article 25 Paragraph C.

[36] MWI. 2009. op.cit. pp. 45.

[37] Ministry of Water and Irrigation. January 2015. loc.cit.

[38] MercyCorps. 2014. Tapped out: Water scarcity and refugee pressures in Jordan. Amman: Mercy Corps. Available at: https://www.mercycorps.org/sites/default/files/MercyCorps_TappedOut_Jordan WaterReport_March204.pdf. Last Accessed (13.04.2016).

[39] Loewe, M., Blume, J., Schönleber, V., Seibert, S., Speer, J. and Voss, C. 2007. Op.cit.

Bottled water has an established market with local businesses collecting, filtering and storing water in 20 litre jerry cans.[40] Bottled and tank water can be 20–46 times more expensive than piped water[41] and are not considered 'improved water sources' by WHO standards.[42]

On water quality, the GoJ claims that 98 percentage of this water complies with Jordanian standards (based on WHO standards).[43] However, Oxfam argues that households store water for an extended period of time in unhygienic conditions, leading to contamination.[44] Jan Schafer[45] describes how many households are concerned over ingesting piped water following an instance in 2007 where a contaminated water supply led to hundreds of citizens falling sick with fever and diarrhoea.[46] Since then, public perceptions of piped water quality have remained low, with 80 percentage of the interviewed water consumers in Amman considering piped water to be contaminated.[47]

Overall, high water scarcity has led to a securitization of water resources. As a result, Jordan's water institutions retain strong central control over its limited resources. Water provision in Jordan has developed in such a way that household water access is primarily dependent on storage capacity and purchasing power. Intermittency and poor perceptions of piped water quality have led water needs to be sourced from private suppliers. As this type of water is more expensive per unit volume,[48] poorer households are structurally vulnerable. Richer households can store larger sums of water and have more purchasing power to supplement water needs through the private market.

Syrian refugees in Jordan

The UNHCR has registered 638,633 Syrian refugees in Jordan in 2016.[49] The real number is much higher as many choose not to register.[50] Refugees may register with UNHCR and reside in one of five refugee camps.[51] However only 17 percentage of refugees reside within these camps. Non-camp refugees are located primarily in the

40 Wildman, T. 2013. op.cit pp. 22.
41 Ibid. pp. 10.
42 UNICEF and WHO. 2006. Core questions on drinking water and sanitation for household surveys. pp. 16. Available at: http://www.who.int/water_sanitation_health/monitoring/oms_brochure_core_questionsfinal24608.pdf (Last Accessed: 12.02.2016).
43 MWI. 2009. op.cit. pp. 36.
44 Laurence Hamai, Ruth McCormack, Eva Niederberger, Lou Lasap. March 2013. op.cit. pp. 12. https://www.alnap.org/help-library/integrated-assessment-of-syrian-refugees-in-host-communities-jordan-emergency-food.
45 Schäfer, P.J. 2013. Chapter 3: Water Security. pp. 40. *In*: Philip Jan Schafer (ed.). Human and Water Security in Israel and Jordan. Weisenheim am Berg Germany: Springer.
46 Schäfer, P.J. 2013. Chapter 3: Water Security. pp. 40. *In*: Philip Jan Schafer (ed.). Human and Water Security in Israel and Jordan. Weisenheim am Berg Germany: Springer.
47 Potter, Robert B. Darmame, Khadija Nortcliff, and Stephen. 2010. op.cit.
48 Wildman, T. 2013. op.cit. pp. 20.
49 UNHCR. 19 Apr. 2016. Syria Regional Refugee Response (Jordan Overview), Available at: http://data.unhcr.org/syrianrefugees/country.php?id=107 (Accessed: 19.14.2016).
50 Berti, B. 2015. The Syrian Refugee Crisis: Regional and Human Security Implications. *Strategic Assessment* 17(4): 41.
51 Francis, A. 2015. Jordan's Refugee Crisis. Carnegie Endowment for International Peace. Available at: http://carnegieendowment.org/files/CP_247_Francis_Jordan_final.pdf (Last Accessed: 15.02.2016) pp. 1.

northern and central governorates. The northern governorates of Amman, Irbid and Mafraq host more than 76 percentage of refugees.[52] Urban centres are particularly saturated, with refugees constituting 52 percentage of inhabitants in Mafraq.[53]

WASH services in refugee camps cover basic water needs but at a very high operational and environmental costs.[54] Furthermore, the constricting nature of refugee camps causes many refugees to reside outside them.[55] REACH estimates that 75 percentage of non-camp Syrians live in rented accommodation, although the housing crisis means that many of these rented units are sub-standard.[56]

Refugees not in rented accommodation live in informal settlements with no access to piped water. Moreover, as most Syrians cannot work legally, many refugees work illegally or rely on asset-depleting coping strategies in order to fund water needs.[57]

Private vendors consist of informal networks of privately owned trucks. Private vendors also tend to discriminate against refugees as native households have more social capital and receive preferential access through social and familial ties. As truck drivers are almost exclusively male,[58] this can have a compounded effect against female refugees who are widowed or unmarried.[59]

Refugee Impact

At 140 m³ per capita, according to the FWSI, pre-crisis water scarcity was at a level constituting a "major constraint to human life".[60] A refugee influx equivalent to 10 percentage of the population[61] will undoubtedly strain water provision, further decreasing water per capita. According to MercyCorps, the average water supply of the hardest hit communities has fallen to under 30 litres/day.[62] In many northern governorates, Syrian refugees have doubled the demand for water.[63]

---

[52] Stave, E. and Hillesund, S. 2015. Impact of Syrian refugees on the Jordanian labour market: Findings from the governorates of Amman, Irbid and Mafraq, Jordan: International Labour Organization Regional Office for the Arab States and Fafo pp. 14.

[53] Ibid. pp. 4.

[54] UNHCR, UNICEF and WFP. January 2014. Joint Assessment Review of the Syrian Refugee Response in Jordan, Available for download at: https://data.unhcr.org/syrianrefugees/download.php?id=4309. Last Accessed: (15.04.2016) pp. 19.

[55] Berti, B. 2015. Op.cit. pg. 44.

[56] REACH. January 2014. Syrian Refugees in Host Communities Key Informant Interviews/District Profiling: Available online at: http://www.reach-initiative.org/wp-content/uploads/2014/03/lea.macias-18022014-093340-REACH-BritishEmbassyAmman_Syrian-Refugees-in-Host-Communities_Key-Informant-Interviews-and-District-Profiling_Jan2014.pdf (Last accessed: 19.04.2016). pp. 12.

[57] UNHCR, WFP and UNICEF. 2015. Vulnerability Assessment of Syrian Refugees in Lebanon, pp. 56. Available at: https://data.unhcr.org/syrianrefugees/download.php?id=10006 (Last Accessed: 24.03.2016).

[58] MercyCorps. 2014. Op.cit pp. 19.

[59] Wildman, T. 2013. loc.cit.

[60] Schäfer, P.J. 2013. Chapter 3: Water Security. pp. 20. *In*: Philip Jan Schafer (ed.). Human and Water Security in Israel and Jordan. Weisenheim am Berg Germany: Springer.

[61] Calculated from pre-crisis population (6,460,000) and Syrian refugee population (638,633). Source: World Bank Data (2016) Population, total. Available at: http://data.worldbank.org/indicator/SP.POP.TOTL (Accessed: 02.04.2016).

[62] MercyCorps. 2014. Op.cit pp. 4.

[63] Francis, A. 2015. op.cit. pp. 16.

The SRC has exacerbated the structural vulnerabilities within the Jordanian population predating the crisis. Although water holds intrinsic socio-cultural value, access to water has become commodified to the extent that it is contingent on one's relationship to the cash-nexus. Since a significant determinant of water access is purchasing power, the economic impacts of the SRC should be considered.

An increase in the demand for accommodation and falling supplies causes rent inflation, forcing households to divert a larger proportion of their income to shelter. In 2013, rent across the five northernmost governorates averaged between 125–175 JOD/month,[64] a price increase from 50 JOD/month in 2012. In the northern towns of Mafraq and Ramtha, average total prices tripled, with some rental prices rising six-fold.[65]

MercyCorps reveals that many Jordanians feel that Syrians accept employment at below pre-crisis market-rates, contributing to wage depression. However, Syrians argue that they act in desperation due to restrictive labour laws and exploitative Jordanian employers.[66] The International Labour Organization (ILO) reported that unemployment grew from 14.5 to 22 percentage between 2011 and 2014.[67]

The degree to which worsening economic conditions are exclusively a result of refugees is a subject of debate. The effects of regional instability on trade and tourism also needs to be examined. Nevertheless, the FAFO institute for Labour and Social Research identifies a causal relationship between Syrian refugees and deterioration in wages and working conditions.[68]

In the existing institutional framework this deterioration contributes to an increase in household water insecurity through a reduction in the disposable income available to supplement household water deficits. This disproportionately affects poorer households through decreasing water affordability. UNESCO reports that 57 percentage of individuals living under the poverty line in 2008 came from three governorates: Amman, Zarqa and Irbid.[69] These coincide with those governorates hosting the most refugees.

The resulting water insecurity manifests in many ways. For example, households struggling to cover water deficits may resort to water consumption strategies. This may mean prioritizing water for drinking often at the expense of water for hygiene and sanitation.[70]

---

[64] Laurence Hamai, Ruth McCormack, Eva Niederberger, Lou Lasap. March 2013. op.cit pp. 18. https://www.alnap.org/help-library/integrated-assessment-of-syrian-refugees-in-host-communities-jordan-emergency-food.

[65] Doa Ali, Hussam Da'anah, Sara Obeidat, Mohammad Hijazi and Sufian Ahmad. 2015. A Town's Sudden Growth Jordanians and Syrians Share the Poverty of Mafraq. Available at: http://7iber.com/2014/08/a-towns-sudden-growth/ (Accessed: 02.04.2016): As cited in Francis, A. (2015) Jordan's Refugee Crisis. Carnegie Endowment for International Peace pp. 16. Available at: http://carnegieendowment.org/files/CP_247_Francis_Jordan_final.pdf (Last Accessed: 15.02.2016).

[66] MercyCorps. October 2012. Analysis of Host Community-Refugee Tensions in Mafraq, Jordan. Available for download: https://data.unhcr.org/syrianrefugees/download.php?id=2958 (Last Accessed: 20.03.2016).

[67] Stave, E. and Hillesund, S. 2015. op.cit. pp. 46.

[68] Ibid. pp. 115.

[69] UNDP. January 2013. Jordan Poverty Reduction Strategy Final Report, pp. 128. Available at: http://planipolis.iiep.unesco.org/upload/Jordan/Jordan_PRSP_2013.pdf (Last Accessed: 15.04.2016).

[70] Wildman, T. 2013. op.cit pp. 8.

Lebanon—Water Sector

Water resources

The Republic of Lebanon enjoys relative water abundance. Emmanuelle Kunigk describes Lebanon as "un chateau d'eau".[71] Surface-water makes up 80 percentage of total water resources.[72] There are 40 major streams and 12 perennial rivers originating in the western mountain peaks and flowing into the Mediterranean. The Litani River, Lebanon's longest river, falls entirely within Lebanese territory.[73] The El-Kebir River, along with the Hasbani and Assi-Orontes, flow internationally.

Groundwater sources include eight major aquifers, recharged by rainfall and snowmelt seepage. In 2011, water averaged 1000 m³ per capita.[74] According to UNESCO, Lebanon experiences water stress, but not water scarcity.[75] Although Lebanon is relatively water-rich, yet several factors limit water supply.

Firstly, water supply is seasonally and geographically variable. Although the river basins of the Litani, the Hasbani and the Assi-Orontes cover 45 percentage of the country,[76] over 50 percentage of Lebanon is subject to desertification.[77] Ninety percentage of total precipitation falls in the winter season (November–April). There is limited precipitation in the summer season, with several areas receiving no rainfall. This implies the need for water storage to ensure continuous water supply.[78] In reality, 50 percentage of total precipitation is lost through evapotranspiration and capturing water on a large-scale in the coastal region (containing the majority of surface water flow) can be difficult due to dense urbanisation.[79]

Secondly, Lebanon's water infrastructure is severely underdeveloped. Lebanon's civil war (1975–1991)[80] and two decades of Israeli occupation have considerably damaged water infrastructure. This has contributed to the lack of existing storage facilities—with Lebanon only capable of storing 6 percentage of its water resources.[81]

Lastly, water institutions are constricted by a political culture based on "sectarian and patron-client consideration".[82] This culture has contributed to a fundamental

---

[71] Kunigk, E. 1998/1999. op.cit. pp. 1.

[72] UNHCR. January 2014. Lebanon WASH sector: risk of water shortages this summer Beirut, 26th January 2014 pp. 1. Available to download at: https://data.unhcr.org/syrianrefugees/download.php?id=4630 (Last Accessed: 12.03.2016).

[73] FAO AQUASTAT. 2008. Lebanon: Geography, Climate and Population, Irrigation in the Middle East region in figures—AQUASTAT Survey 2008, Water Report 34 pp. 5. Available at: http://www.fao.org/nr/water/aquastat/countries_regions/lbn/LBN-CP_eng.pdf (Accessed: 12.04.2016).

[74] Connor, R. 2015. The United Nations world water development report 2015: water for a sustainable world (Vol. 1). UNESCO Publishing. pp. 78.

[75] UNESCO. 2012. opt.cit. pp. 124.

[76] FAO AQUASTAT. 2008. opt.cit pp. 5.

[77] Karen Assaf, Bayoumi Attia, Ali Darwish, Batir Wardam and Simone Klawitter. 2004. op.cit. pp. 159.

[78] Republic of Lebanon–Ministry of Environment. 2001. State of the Environment Report. Chapter 8 (Water). pp. 189. Available for download: http://www.moe.gov.lb/getattachment/The-Ministry/Reports/State-Of-the-Environment-Report-2001/Chap-8-Water.pdf.aspx: (Last Accessed: 20.03.2016).

[79] Ibid.

[80] Karen Assaf, Bayoumi Attia, Ali Darwish, Batir Wardam and Simone Klawitter. 2004. op.cit. pp. 150.

[81] Armstrong, M. Thursday 5 March 2015. Can Lebanon implement much-needed water reform? Available at: http://www.middleeasteye.net/in-depth/features/lebanon-capable-implementing-much-needed-reform-its-water-sector-187038597 (Accessed: 13.04.2016).

[82] Makdisi, K. 2008. op.cit, pp. 162.

inequity in water access and an over-exploitation of water resources.[83] In 2010, the World Bank predicted that Lebanon would face chronic water shortages by 2020.[84]

Water governance

In 2000, Law 221/2000 (the Water Law) delegated the Ministry of Energy and Water (MEW) as the highest water authority, holding guardianship over four regional Water Authorities (WA).[85] These WA's hold regional financial and administrative autonomy over water provision.[86]

Lebanese water has been historically mishandled.[87] Before the Water Law, no specific legislation existed regarding municipal water, with surface and groundwater perceived as "public domain".[88] The water sector suffers from a lack of policy co-ordination due to the countries ethnically and religiously fractured populace. Former MEW Minister Elie Hobeika declared, "Everything in the Lebanese water sector is down to politics".[89]

This political capture has led WA's to suffer technically and financially. In 2010, cost efficiency in the WA's averaged only 47 percentage[90]—being 11 percentage for the WA for Beka'a.[91] Over 50 percentage of pipes that transfer and distribute water have passed their useful-life,[92] with national leakage loss at 35–50 percentage.[93] Infrastructure development often depends on investment from public-private partnerships.[94] This is highlighted in the MEW's 2010 National Water Sector Strategy (NWSS), which involves a ten-years strategy to increase private-sector participation.[95] The NWSS strongly espouses a water productivity paradigm, transforming many WA's into commercial institutions to increase cost efficiency.

Socially, water is perceived to be a common heritage,[96] or a public good.[97] Gebran Bassil, from the MEW, describes water "as a right for every citizen, a resource for

---

[83] Karen Assaf, Bayoumi Attia, Ali Darwish, Batir Wardam and Simone Klawitter. 2004. op.cit. pp. 160.

[84] World Bank. 2010. Republic of Lebanon Water Sector: Public Expenditure Review, Report No. 52024-LB Sustainable Development Department Middle East and North Africa Region pp. i. Available at: http://www-wds.worldbank.org/external/default/WDSContentServer/WDSP/IB/2015/08/31/090224b0828b4948/1_0/Rendered/PDF/Republic0of0Le0c0expenditure0review.pdf (Last accessed: 22.03.2016).

[85] Hamamy, G. 2007. EGM on the production of statistics on natural resources and environment. Presidency of the Council of Ministers. Central Administration of Statistics.

[86] Makdisi, K. 2008. op.cit, pp. 167.

[87] Karen Assaf, Bayoumi Attia, Ali Darwish, Batir Wardam and Simone Klawitter. 2004. op.cit. pp. 156.

[88] Kunigk, E. 1998/1999. opt.cit. pp. 12.

[89] Kunigk, E. 1998/1999. opt.cit. pp. 17.

[90] World Bank. 2012. Lebanon Country Water Sector Assistance Strategy 2012–2016, Report No. 68313-LB Sustainable Development Department Middle East and North Africa Region pp. 2.

[91] World Bank. 2010. op.cit. pp. i.

[92] Eng Gebran Bassil. December 2010. National Water Sector Strategy, Ministry of Energy and Water Slides slide 8 retrieved from: http://planipolis.iiep.unesco.org/upload/Jordan/Jordan_PRSP_2013.pdf (Last Accessed: 13.04.2016).

[93] FAO AQUASTAT. 2008. opt.cit pp. 5.

[94] World Bank. 2010. op.cit. pp. 13.

[95] Eng Gebran Bassil. December 2010. op.cit. 21.

[96] FAO AQUASTAT. 2008. opt.cit pp. 13.

[97] Catafago, S. 2005. Restructuring water sector in Lebanon: Litani river authority facing the challenges of good water governance. pp. 75. *In*: Hamdy, A. and Monti, R. (eds.). Food Security Under Water Scarcity in the Middle East: Problems and Solutions. Bari: CIHEAM.

the whole country".[98] Citizens feel entitled to groundwater resources without state authority. The accompanying corruption and lack of enforcement has led to gross over-exploitation, with private water suppliers contributing to 30–40 percentage of total household water demand.[99]

Water provision

According to the World Bank, 99 percentage of Lebanese households have access to an improved water source.[100] The CAS census on housing and establishments reports that 79 percentage of buildings were connected to water supply networks. The highest rates of connection were recorded in the capital, Beirut (93 percentage), a wealthy, relatively populous settlement. The lowest rates of connection were recorded in Hermel (41 percentage), a poorer, more rural settlement.[101]

Household connectivity also correlates with income, with 38 percentage of Lebanese households in the lower income quintile remaining unconnected compared to 14 percentage in the highest quintile.[102] Karim-Philip Eid-Sabbagh argues that this relates less to the price of connections and more to the regional disparities in water infrastructure. Many households remain unconnected to the water network because no regional connection exists. This mirrors patterns of investment in water infrastructure, with poorer and marginalised communities receiving less infrastructure funding than richer and influential communities.[103]

The MEW subsidizes the water sector with tariffs set by WA's in order to recuperate costs. A flat rate tariff is set which varies between WA's.[104] The WA for North Lebanon is sourced by gravity distribution from a nearby aquifer. This is cheaper than pumping water from underground or long distances,[105] such as for Beirut. These costs are reflected in tariff prices for networked water, linking water affordability, and hence access.

Due to geographical variability, Lebanon suffers from seasonal water shortages.[106] Lebanon cannot guarantee sufficient household water for all of its citizens all year round. As a result, water supply is effectively intermittent.[107] The majority of

---

[98]   Eng Gebran Bassil. December 2010. op.cit. 1.

[99]   Karen Assaf, Bayoumi Attia, Ali Darwish, Batir Wardam and Simone Klawitter. 2004. op.cit. pp. 156.

[100] World Bank Data. 2016. Improved water source (% of population with access). Available at: http://data.worldbank.org/indicator/SH.H2O.SAFE.ZS (Accessed: 28.03.2016).

[101] Republic of Lebanon–Ministry of Environment. 2001. op.cit. pp. 116.

[102] Eid-Sabbagh and Karim-Philipp. 2015. A political economy of water in Lebanon: water resource management, infrastructure production, and the International Development Complex. PhD Thesis. SOAS, University of London pp. 195.

[103] Ibid.

[104] World Bank. 2012. op.cit pp. 4.

[105] FAO AQUASTAT. 2008. opt.cit pp. 14, and Catafago, S. 2005. loc.cit.

[106] Darwish, A. 2004. Water as a human right: Assessment of water resources and water sector in Lebanon. Available at: ttps://www.boell.de/sites/default/files/assets/boell.de/images/pics_de/internationalepolitik/GIP11_Lebanon_Ali_Darwish.pdf (Last accessed: 10.02.2014).

[107] World Bank. March 2009. LEBANON Social Impact Analysis–Electricity and Water Sectors, Social and Economic Development Group Middle East and North Africa Region: Available at: http://www.pseau.org/outils/ouvrages/world_bank_lebanon_social_impact_analysis_electricity_and_water_sectors_2009.pdf (Last Accessed: 13.04.2016).

households do not receive a continuous water supply.[108] For example, during the summer season the Beirut-Mount Lebanon region (accounting to 60 percentage of total connected households) receives only three hours of daily water supply.[109]

Households supplement water deficits through private sources.[110] These include artesian wells, service water trucks or bottled drinking water.[111] According to the NWSS, there are roughly 43,000 private wells.[112] Furthermore, given that up to 75 percentage of water expenditure goes to private suppliers,[113] the majority of water provision can be considered as privatised. The resulting water commodification impacts access along income lines, with private water costing three times as much as network water.[114]

Lebanon is divided into 1108 municipalities (LCPS, 2015). Eight hundred of which are considered administratively and fiscally weak (Atallah, 2012), incapable of covering local water demands. Eid-Sabbagh notes:

> "As public water services are completely absent in some areas, or partially available, local committees organise their own services and rely on the provision of water through political parties or other patrons. These clientalist networks are configured according to very local political dynamics and reflect the socio-political history of the specified areas."[115]

Municipalities can join one of the fifty one municipal unions—divided along political-demographic lines. Unions enjoy the benefits of sharing pooled resources, collective decision-making and greater lobbying power for regional development.[116]

On water quality, 50–75 percentage of all Lebanese water sources do not fulfil WHO standards for bacterial contamination.[117] Uncontrolled sewage disposal is a substantial obstacle to clean water.[118] The World Bank described wastewater sector development at an "embryonic stage".[119] Nearly all of Lebanon's domestic wastewater

---

[108] World Bank. December 23rd, 2013. Feature Story: Water in Lebanon: Matching Myth with Reality, Available at:http://www.worldbank.org/en/news/feature/2013/12/23/water-in-lebanon-matching-myth-with-reality (Accessed: 12.03.2016).

[109] World Bank. 2010. op.cit. pp. i.

[110] Makdisi, K. 2008. op.cit, pp. 169/170.

[111] Ecodit. 2015. Strategic Environmental Assessment for the New Water Sector Strategy for Lebanon, Regional Governance and Knowledge Generation Project: pp. 32. Available at: http://www.moe.gov.lb/getattachment/5ac7c2d8-4a3a-4460-a6c8-daa8acd42f1b/STRATEGIC-ENVIRONMENTAL-ASSESSMENT-FOR-THE-NEW-WAT.aspx (Last Accessed: 09.04.2016).

[112] Eng Gebran Bassil. December 2010. op.cit. 39.

[113] Eng. Ahmad Nizam. 7,8 December 2011. Water Sector Reform in Lebanon and Impact on Low Income Households, 4th ACWUA Best Practices Conference Water and Wastewater Utilities Reform "Changes and Challenges": pp. 10. Slides retrieved from: http://www.acwua.org/sites/default/files/ahmad_nizam.pdf (Last Accessed: 12.04.2016).

[114] World Bank. 2012. op.cit. pp. 2.

[115] Eid-Sabbagh and Karim-Philipp. 2015. op.cit. pp. 200.

[116] MercyCorps. March 2014. Policy Brief: Engaging Municipalities in the Response to the Syria Refugee Crisis in Lebanon. March 2014. pp. 3. Available to download from: http://www.alnap.org/pool/files/mercy-corps-lebanon-policy-brief-engaging-municipalities-(english).pdf (Last Accessed: 04.04.2016).

[117] Eid-Sabbagh and Karim-Philipp. 2015. op.cit. pp. 195.

[118] Republic of Lebanon–Ministry of Environment. 2001. op.cit. pp. 117.

[119] World Bank. 2010. op.cit. pp. 12.

is discharged directly into nearby watercourses, resulting in health issues as those same water sources are widely used for domestic purposes.[120] Information concerning private water quality is speculative due to lack of monitoring. Although a 2002 UN study indicates that 24 percentage of sampled bottled water was microbiologically contaminated.[121]

Overall, although water is politicized, low institutional capacity has weakened state oversight of the water sector.[122] Access to water networks depends on geographical residence, in turn affected by social-political legacies and infrastructure development. The inability of the WAs to meet basic household water needs has led to the dominance of a private-water market. The resulting commodification of water discriminates along socio-economic lines. Water quality is also a pressing concern, with wastewater management severely under-developed.

Syrian refugees in Lebanon

The population of Lebanon is approximately 4.54 million.[123] According to the UNHCR, there are 1,055,984 Syrian Refugees,[124] while in Jordan; the real number is likely to be higher.[125] Unlike Jordan, no government refugee camps exist to accommodate refugees in Lebanon. Domestic political realities have led to a no-camp policy.[126] Lebanon has historically been a safe haven for regional forced migration, holding the highest per-capita concentration of refugees worldwide with refugees constituting 30 percentage of its population.[127]

Lebanon's refugee population is widely dispersed (Fig. 13.2), with urban centres in Akkar, North Lebanon and Mount Lebanon being the most densely packed governorates. The UNHCR reported in 2015 that 30 percentage of the refugee population lack access to safe drinking water.[128] Fifty-seven percentage of Syrian

---

[120] World Bank. 2013. Lebanon: Economic and Social Impact Assessment of the Syrian conflict. pp. 110. Available at: http://www.undp.org/content/dam/rbas/doc/SyriaResponse/Lebanon%20Economic%20 and%20Social%20Impact%20Assessment%20of%20the%20Syrian%20Conflict.pdf (Last Accessed: 09.04.2016).

[121] UNDP. December 2013. Lebanon Millennium Development Goals Report 2013. Available at: http:// www.undp.org/content/dam/undp/library/MDG/english/MDG%20Country%20Reports/Lebanon/ MDG%20English%20Final.pdf. As cited *In*: Biswas, A., Rached, E. and Cecilia Tortajada (eds.). Water as a Human Right in the Middle East and North Africa. New York: Routledge Taylor and Francis Group, pp. 171.

[122] World Bank. 2010. op.cit. pp. 8.

[123] World Bank Data. 2016. Population, total. Available at: http://data.worldbank.org/indicator/SP.POP. TOTL (Accessed: 02.04.2016).

[124] UNHCR. 19 Apr. 2016. Syria Regional Refugee Response (Jordan Overview), Available at: http://data. unhcr.org/syrianrefugees/country.php?id=107 (Accessed: 19.14.2016).

[125] Berti, B. 2015. Op.cit pp. 41.

[126] Berti, B. 2015. Op.cit pp. 41.

[127] UNHCR and the Government of Lebanon. 2015. Lebanon Crisis Response Plan 2015–2016 pp. 41. Available at: http://www.3rpsyriacrisis.org/wp-content/uploads/2016/04/3RP-2015-Annual-Report.pdf (Last Accessed: 11.04.2016).

[128] UNHCR. March 2015. Refugees from Syria: Lebanon pp. 13. Available to download at: https://data. unhcr.org/syrianrefugees/download.php?id=8649 (Last Accessed: 10.03.2016).

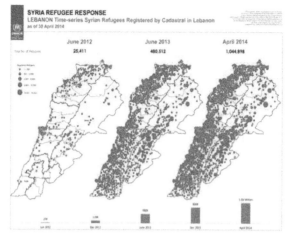

Figure 13.2 Distribution of Syrian refugees in Lebanon (UNHCR, 2014).

Color version at the end of the book

refugees live in rented accommodation very similar to Lebanese households,[129] with comparable water network access. Sixteen percentage of refugees live in informal settlements characterised by poor or no access to water or sanitation, representing some of the most water-insecure.[130]

Refugee impact

Poor regulation and an uncontrolled private water sector have led to the gross over-exploitation of groundwater resources, with pre-crisis rates exceeding the renewable threshold.[131] Forced migration has increased water demands and negatively influences the situation.

Chronic water shortages have resulted in a dependency on private water sector, which are unregulated and unsubsidized. In Jordan, reduced income-generation and disposable income will have contributed to household water insecurity. In the Bekaa valley, a 60 percentage wage reduction has been reported, similar to North Lebanon.[132] The population increase has raised rental prices.[133] In addition to this, the World Bank calculated that the unemployment rate would double, pushing 170,000 Lebanese citizens into poverty.[134]

The refugee crisis has compounded a poor pre-crisis economy, with a high structural unemployment and deficit. The crisis will only act to further reduce funds

---

[129] Norwegian Refugee Council. June 2014. A precarious existence: the shelter situation of refugees from Syria in neighbouring countries. Pg. 5. Available here: http://www.nrc.no/arch/_img/9179861.pdf (Last Accessed: 13.03.2016).

[130] Ibid. pp. 6.

[131] UNHCR. January 2014. loc.cit.

[132] Vliet and Hourani. 2014. loc.cit.

[133] ILO. 2013. Assessment of the impact of Syrian Refugees in Lebanon and their employment profiles pp. 38. Available at: http://www.ilo.org/wcmsp5/groups/public/---arabstates/---ro-beirut/documents/publication/wcms_240134.pdf (Last Accessed: 19.04.2016).

[134] World Bank. 2013. op.cit. pp. 65.

the MEW can allocate to WAs. In 2016, the MEW appealed for over $391 million to deal with the crisis.[135] Heads of municipalities expressed that they are unable to provide for the greater demand for water and wastewater management.[136] A 2014 MercyCorps assessment amongst 12 municipalities found that 71 percentage of those surveyed indicated that conditions had worsened. Moreover, 79 percentage of municipalities identified water service provision as their greatest challenge.[137]

Due to regional inequalities, the impact of the refugee influx varies across the country. The most affected Lebanese households are located in underdeveloped areas such as North Lebanon and Bekaa, which are characterised as the poorest and least developed.[138] Water quality continues to be a threat to a clean water supply. Pre-crisis, 92 percentage of Lebanon's wastewater ran untreated into surrounding watercourses.[139] With government funds being allocated to dealing with refugees, there has been little improvement in wastewater management. Between 2012–2014, the rising refugee population produced an additional 8 percentage–14 percentage of wastewater,[140] suggesting that refugees have only aggravated this threat.

Comparison and Analysis

Comparison

Due to their proximity to Syria, Jordan and Lebanon are among the states most affected by the SRC. Both countries have received a significant number of refugees, with detrimental impacts on water security. For households, both water accessibility and water availability have been stressed through a number of means.

Despite tribal influences over water resources and *wasta* corruption, the GoJ maintains authority over water provision. Severe water scarcity has mobilized the government to be proactive on water issues, implementing a strict water rationing system in order to manage demand. In Lebanon, seasonal variability and poor infrastructure have undermined state provision of domestic water, leading to persistent water shortages. An inadequate institutional and regulatory capacity has resulted in a lack of state control over water resources. The Lebanese population is distinctively sectarian and far more heterogeneous than Jordan. The tendency to decentralize water provision has exacerbated regional inequalities. Because of these institutional differences, the SRC has had differing levels of impact on water affordability, water quality and water availability.

---

[135] UNHCR. 2016. Regional Refugee and Resilience Plan 2016–2017 in response to the Syria Crisis pg. 13. http://www.3rpsyriacrisis.org/wp-content/uploads/2015/12/3RP-Regional-Overview-2016-2017.pdf (Last Accessed: 19.04.2016).

[136] MercyCorps. March 2014. op.cit. pp. 10.

[137] Ibid pp. 3.

[138] Vliet and Hourani. 2014. Regional differences in the conditions of Syrian refugees in Lebanon. Available at: http://cskc.daleel-madani.org/paper/regional-differences-conditions-syrian-refugees-lebanon (Last Accessed: 09.04.2016).

[139] UNHCR and the Government of Lebanon. 2015. op.cit pp. 47.

[140] EU. 2014. Lebanon Environmental Assessment of the Syrian Conflict & Priority Interventions pp. 6. https://www.aub.edu.lb/facilities/chsrm/Documents/EASC-ExecutiveSummaryEnglish.pdf (Last Accessed: 13.04.2014).

Due to water's necessity for sustaining livelihoods, state subsidies in both Jordan and Lebanon have made public water largely affordable. However, as both countries have intermittent water supplies, wealthier households are able to cope better through enhanced storage capacities and better connectivity—this is more prevalent in Jordan where rationing is more severe. High dependency on costly private water in both countries also discriminates along socio-economic lines, particularly more so in Lebanon where 75 percentage of water consumption is supplied by non-state actors.[141]

Both countries have experienced rises in unemployment, rent and private water prices because of the influx of refugees. Jordanian and Lebanese households find that they have less disposable income to supplement water deficits, along with a lower capacity for income-generation. For poorer households this may mean having to prioritize drinking over hygiene.[142] Overall, water has become less affordable in both countries, and hence less accessible to less affluent households.

The GoJ manages wastewater collection in Jordan effectively, with programs to reuse treated wastewater for agricultural demands.[143] The largest threat to water quality in Jordan remains the environmental costs of refugee camps. This is particularly the case for Za'atari refugee camp, which was developed in haste and directly over an over-exploited aquifer.[144] Water quality in Lebanon is a substantial threat to future household water security. Poor wastewater management risks household health by contaminating main water resources.[145] The addition of wastewater produced by refugees' compounds this threat.

Syrian refugees have reduced general water availability as rations are spread more thinly. Economically, host communities have sustained huge fiscal costs. Addressing the UN Sustainable Development Summit in 2015, the President of Lebanon, Tammam Salam, stated that hosting Syrian refugees has cost Lebanon a third of its GDP.[146] Similarly, in an interview for the BBC, King Abdullah II of Jordan attributed 25 percentage of the Jordan's budget to refugee-related costs.[147] Water infrastructure in both countries requires maintenance—markedly so in the Lebanese context. Increased water demand and wastewater production has come at a time when governments have had fewer resources to fund municipal water services. As such, the state's capacity to mitigate impacts on water security is diminished.

[141] Eng. Ahmad Nizam. 7,8 December 2011. loc.cit.

[142] Wildman, T. 2013. op.cit pp. 8.

[143] Amani Alfarra, Eric Kemp-Benedict, Heinz Hötzl, Nayif Sader, Ben Sonneveld. 2011. A framework for wastewater reuse in jordan: utilizing a modified wastewater reuse index. *Water Resources Management* 25(4): 1153–1167. Available at: http://link.springer.com/article/10.1007%2Fs11269-010-9768-8 (Accessed: 06.01.2016).

[144] Al-Harahsheh, S., Al-Adamat, R. and Abdullah, S. 2015. The impact of za'atari refugee camp on the water quality in amman-zarqa basin. *Journal of Environmental Protection* 6: 16–24. Available at (http://file.scirp.org/pdf/JEP_2015011510265820.pdf) (10.4236/jep.2015.61003) (Last Accessed: 20.04.2016).

[145] World Health Organization. 2006. Guidelines for the Safe Use of Wastewater, Excreta and Greywater: Policy and regulatory aspects (Vol. 1). *World Health Organization.*

[146] Statement by Mr. H.E. Tamaam Salam. 2015. President of the Council of Ministers of the Republic of Lebanon (At the United Nations Sustainable Development Summit) New York, Saturday, September 26, 2015. Available online at: https://sustainabledevelopment.un.org/content/documents/19451lebanon.pdf (Last accessed: 23.04.2016).

[147] BBC Middle East. 2 February 2016. Syria conflict: Jordanians 'at boiling point' over refugees, Available at: http://www.bbc.co.uk/news/world-middle-east-35462698 (Accessed: 09.04.2016).

Long-term water security for households requires a sustainable water supply. Neither country has utilized its water resources sustainably. In Jordan, limited resources constrict effective water management. With increasing water demands, Jordan has been forced to increase rates of groundwater extraction,[148] which already exceeded renewable yields. Lebanon's groundwater extraction is reminiscent of Hardin's 'Tragedy of the Commons',[149] where unregulated private extraction threatens the survival of water resources.

Institutional analysis

The institutional analysis in the discussion above reveals that Jordan's water scarcity has influenced the development of strong water institutions. The GoJ's comparatively strong authority over water resources, along with a cohesive water sector has afforded it more robust water governance. On the other hand, Lebanon's history of domestic conflict has damaged its water infrastructure. An ethnically and politically fractured populace has led to a deeply heterogeneous water sector that, along with geographical variability, has weakened Lebanon's capacity to provide household water needs. As a result, Lebanon has comparatively weaker water governance.

Clearly, each countries pre-crisis political and socioeconomic context equips in a different way to deal with the influx. Despite an unarguably negative impact on water security in both cases, a comparison between the two suggests that Lebanon's low institutional capacity affords it less room for coping with the crisis compared to Jordan's state-centric control. This conclusion is reached in spite of Lebanon's relative water abundance—reinforcing the importance that institutions play in maintaining water security. The World Resources Institute prediction that by 2040 Lebanon will be more water stressed than Jordan supports this finding.[150]

In suggesting that water institutions are a result of water scarcity, it is also important to consider whether the SRC itself is endogenous to rising water scarcity in the region. The Proceedings of The National Academy of Sciences considers the central role of climate change in the Syrian civil war.[151] Peter Gleick argues that the 2006–2011 regional drought played a causal role in disaffecting rural Syrians.[152] The resulting water insecurity fuelled civil unrest, as livelihoods could not be sustained. This stance does not ignore the influence of the Arab-Spring in inspiring regime

---

[148] International Committee of the Red Cross. March 2015. loc.cit.

[149] Hardin, G. 1968. The tragedy of the commons. *Science* 162(3859): 1243–1248.

[150] WRI. 2015. Water Stress by Country. Available here: http://www.wri.org/sites/default/files/uploads/water_stress_table_large.jpg (Last Accessed: 16.03.2016). As Cited *In*: Maddocks, A., Young, R.S. and Paul, R. August 26, 2015. Ranking the World's Most Water-Stressed Countries in 2040, Available at: World Resources Institute http://www.wri.org/blog/2015/08/ranking-world%E2%80%99s-most-water-stressed-countries-2040 (Accessed: 16.03.2016).

[151] Kelley, C.P., Mohtadi, S., Cane, M.A., Seager, R. and Kushnir, Y. 2015. Climate change in the Fertile Crescent and implications of the recent Syrian drought. *Proceedings of the National Academy of Sciences* 112(11): 3241–3246. Available at: http://www.pnas.org/content/112/11/3241.abstract (Last Accessed 20.03.2016).

[152] Gleick, P.H. 2014. Water, drought, climate change, and conflict in Syria. *Weather, Climate, and Society* 6(3): 331–340.

change. Instead, it critiques the assumption that Syrian refugees have played an exclusively causal role in deteriorating water security. By suggesting that rising water insecurity has influenced the forced migration in question, the SRC can be examined through a wider lens.

It should be noted that this chapter does not argue that the sole determinant of water institutions is water scarcity. This argument would follow a rational choice explanation and betray the critical approach of this chapter. Along with water scarcity, this chapter has also been careful to consider the socio-historical processes, in the development of water institutions. By taking Syrian refugees as an exogenous shock, additional research could be undertaken to study how water institutions and governance will develop in both countries.

Future water governance can be glimpsed through looking at how both countries have responded. In both Jordan and Lebanon, the refugee influx has provoked securitizing responses from state actors. In 2016, King Abdullah of Jordan described his country at "boiling point".[153] Lebanon's MEW has reactivated its strategic national plan and is in the process of adapting it to the current crisis. State politicians have shown hesitation over structural reform that would assist water access for refugees, such as opening up labour markets or expanding asylum requests, for fear of encouraging long-term residency.[154]

Conventional wisdom suggests that as water insecurity rises, deteriorating socioeconomic conditions would fuel social tensions. Increased reports of intolerance have arisen in both countries as the crisis persists. Many Jordanians are aggravated by Syrians "cavalier attitude to water usage".[155] In Lebanon, Syrian refugees exacerbate sectarian divides with negative impacts on water policy cohesion.[156]

There is also evidence to temper these concerns. In Jordan's Mafraq governorate where Syrian refugees have doubled the population, WASH partners rehabilitated and upgraded water and sanitation systems for 500,000 people.[157] Whilst in Lebanon, MercyCorps reported that of those surveyed, over, 65 percentage of Syrians feel "welcomed by the local community" and 85 percentage of Lebanese indicated they "had a moral duty to host refugees".[158]

---

[153] BBC Middle East. 2 February 2016. loc.cit.

[154] Francis, A. 2015. loc.cit.

[155] MercyCorps. May 2013. Mapping of Host Community-Refugee Tensions in Mafraq and Ramtha, JORDAN, pp. 12. www.mercycorps.org: Available to download at: https://data.unhcr.org/syrianrefugees/download.php?id=2962 (Last Accessed: 12.04.2016).

[156] Tan, C. 2015. Strategic sectors, culture and society the syrian refugee crisis: conflicts in the making. pp. 307. *In*: Florensa, S. and Mogherini, F. (eds.). IEMed Mediterranean Yearbook 2015. IEMED Publications. Available at: http://www.iemed.org/observatori/arees-danalisi/arxius-adjunts/anuari/med.2015/IEMed%20Yearbook%202015_Panorama_SyrianRefugeeCrisis_CarolTan.pdf (Last Accessed: 02.04.2016).

[157] UNHCR and the Government of Lebanon. 2015. op.cit. pp. 26.

[158] MercyCorps. June 2013. Things fall apart: Political, economic and social instability in Lebanon. 1–24. Retrieved 10(2013): 2015. pp. 6. Available at: https://www.mercycorps.org.uk/sites/default/files/MC%20Lebanon%20LivelihoodConflict_Assesment_%20Full%20Report%200913.pdf (Last Accessed 04.04.2016).

Conclusion

The Syrian Refugee Crisis has come to dominate international discussion. Widely considered to be the worst humanitarian disaster since the World War II, its impacts on the region have been severe. The context of worsening regional water scarcity provided a compelling reason to study the effects of forced migration on water security within host communities.

The literature review established water security as 'access to safe and sufficient water for domestic needs and sanitation'. Adopting a human-centric perspective, the research methodology took the household as the level of analysis. Relating water security to 'freedom from want', the qualitative indicators of *accessibility* and *availability* were identified in order to guide the assessment.

Adopting a specific definition or perspective, or deciding on a set of indicators, necessitated an omission of other methodologies. Other areas for research include how Syrian refugees have affected the region's hydro-politics, or how changing water demands influence food and energy sectors. Notably, the refugee component emphasizes the urgency of securing basic human needs and the adopted definition reflected this. This is further supported by the acceptance of humanitarian concerns within international water security discourses.

The indicators of accessibility and availability guided a critical institutional analysis of Jordan and Lebanon, studying the political economies surrounding water provision and how they influence access. Many socio-political realities were alluded to in this process, often briefly. As such, further research can be taken to elaborate more thoroughly on effects within countries exclusively as opposed to the comparative methodology utilized here.

The authors have found that the SRC has proved detrimental to water security in both Jordan and Lebanon. Jordan's Minister of Water and Irrigation, Dr. Hazim El-Naser, asserted, "We live with a chronic water problem, and we are now at the edge of moving from a chronic water problem into a water crisis. The element that will trigger this is the number of Syrian Refugees." Fadi Comair, Director General of Hydraulic and Electric Resources, described the Lebanese situation as "truly dramatic".[159]

Comparing the impacts between Jordan and Lebanon produced striking results. Differing levels of water scarcity, along with historical and political factors, have influenced the divergent development of water institutions. Despite Lebanon's relative abundance, impacts on water security have been worse due to poorer institutional characteristics—especially with regards to consequences on water quality. In comparison, the GoJ's authority affords it more room to adapt to the crisis.

Given the scale of the conflict in Syria, refugees are unlikely to be repatriated en masse any time soon. Little more can be requested of host communities or aid-groups. If recommendations are to be made based on this study, they should be directed towards two groups: the international community and water policy practitioners.

---

[159] Whitman, E. 30 May 2013. Refugee influx worsens Jordan's water woes. Available at: http://www.aljazeera.com/indepth/features/2013/05/20135268026616381.html (Accessed: 27.03.2016).

While efforts are focused on tackling the crisis north of the Mediterranean,[160] only 43.1 percentage of the UN's 2015 Syria Response Plan appeal was funded.[161] In dealing with the crisis, international policy has chosen traditional security discourses over human security discourses. The study of water security provides a fresh lens into the neglected outcomes. Indeed, as water scarcity becomes more of a regional concern, water policy practitioners should consider the impacts on basic household needs. In that sense, the SRC has also provided a useful avenue through which to consolidate water security's humanitarian responsibilities towards sustaining livelihoods.

References

Al-Zubari, Waleed K. 2014. *Oil and Water: How can Arabs turn energy into food?* Arab Forum for Environment & Development (AFED).

Al-Zubari, Waleed. 2015. *Arab Environment: Sustainable Water Consumption in Arab Countries.* Arab Forum for Environment & Development (AFED).

Fakih, Ali and Walid Marrouch. 2015. The Economic Impacts of Syrian Refugees: Challenges And Opportunities in Host Countries. *Georgetwon Journal of International Affairs* (Georgetown Journal of International Affairs).

Roudi-Fahimi, Farzaneh, Liz Creel, and Roger-Mark De Souza. 2002. *Finding the Balance: Population and Water Scarcity in the Middle East and North Africa.* Washington, DC: Population Reference Bureau.

---

[160] Middle East Eye. Tuesday 20 October 2015. Lebanon faces water crisis after record winter drought, Availableat: http://www.middleeasteye.net/news/lebanon-faces-water-crisis-after-record-winter-drought-1118861571 (Accessed: 17.04.2016).

[161] Council of the European Union (18/03/2016 Press Release Foreign affairs & international relations) EU-Turkey statement, 18 March 2016, www.mercycorps.org: Available at: http://www.consilium.europa.eu/en/press/press-releases/2016/03/18-eu-turkey-statement/# (Last Accessed: 12.04.2016).

Section 5
Importance of Education

Chapter-14

The 2:1 Resilience Factor, Education for Mitigation and Adaptation

William Humber[1,*] *and Cheryl Bradbee*[2]

INTRODUCTION

The Canadian Prime Minister Justin Trudeau's January 2017 observation that his country needed to phase out fossil fuel production in the oil sands as part of the response to climate change drew a mixed response (Muzyka, 2017). Though grounded in science and fact and acknowledged by some, it was greeted with howls of outrage by others including politicians and citizens. Rhetorically they asked, how could the Prime Minister threaten the well-being of so many? It highlighted an urgent challenge for Canada in upholding the Paris Accord, ratified by Canada on 5 October 2016, while providing a healthy and prosperous future for its citizens.

It demonstrated a large scale ignorance about climate change, the level of urgency, and the necessary responses. Anthropogenic Climate Disruption (ADC) has the potential to challenge everything from water supplies and energy use to food production as well as a host of other quality of life issues. Public commitment to policies required to shift the economy, so that carbon usage is reduced and ecosystems are restored, is essential. Going forward, life will be changed by this issue for which decisions must be made and policies crafted. Prime Minister Trudeau was laying the groundwork for a broader conversation. Canadians need to be informed and aware of the problem, open to imagine its implications for the future, and ready to act individually and corporately.

This chapter examines two aspects of the problem for education around the changes having an impact on Canada. After a brief review of the attitudes that

---

[1] Founding Director, Office of Eco-Seneca initiatives (OESi), Seneca College, 1750 Finch Ave. East, Toronto M2J 2X5.

[2] An educator who has taught at Humber College, Seneca College, Ryerson University, University of Toronto, and York University.
Email: c.bradbee@dal.ca

* Corresponding author: bill.humber@senecacollege.ca

led Canada to this point, we will look at the need to educate all and then offer an approach to formulate content for a new curriculum. We, the authors, want to be clear with our intent. Neither of us are scientists, we are educators. Together we have significant experience teaching in Ontario's colleges and universities. We have experience teaching in programs that do and do not emphasize environmental education. We speak as educators seeking a means to insure that today's students become tomorrow's innovators and involved citizens able to craft solutions for their world and increase our resiliency together.

Evolution of Attitudes in Canada

In addressing the issue of climate change and resilience from an educator's perspective it is worth considering how we got to this point. A quick Canadian history lesson will provide some context.

The 17th century settler arriving in New France, today's province of Quebec, or the United Empire Loyalist of the late 18th century fleeing the new United States of America for the wilds of Upper Canada, today's province of Ontario, encountered forbidding landscapes of trees, swamps, wild animals and other threats. These challenges were interpreted in different ways by these newcomers as opposed to the people who were familiar with and had lived in these landscapes for upwards of ten thousand years.

As late as 1842 the English novelist Charles Dickens, on his tour of North America, wrote, "There was the swamp, the bush, the perpetual chorus of frogs, the rank unseemly growth, the unwholesome steaming earth…" and he meant none of it in a good way (Dickens, 1842).

It was an imperfect world of natural obstacles. Overcoming these was a priority for those with roots largely in Europe for whom advances in comfort, well-being and prosperity were ideals to be pursued. In the case of First Nations peoples they had arrived at an uneasy balance with this imperfect world. In their own practices over ten thousand years or more of residence, First Nations peoples had likewise transformed this world as first encountered. However, these ways were less apparent to the European settlers whose increasingly sophisticated technology made for a more radical transformation.

For the new settler, whether of French or British extraction, or possibly a one-time free African fleeing slavery in the American states, trees were an impediment to farming and their removal required back-breaking labor. Swamps were wasted spaces in which pests and mosquitoes, carrying diseases including malaria, might be found. Wild animals from bears to wolves threatened the very life of an unarmed pioneer.

The elimination of vegetation, the draining of swamps and the killing of those bears and wolves was a priority. Trees were burned for potash, underground aquifers were tapped for water, and rivers were dammed to build mills to produce flour for bread. Such measures reflected a bias for tackling and subduing nature in contrast to the more balanced relationship achieved over the long-term by First Nations peoples. Few would want to undo the consequent raising of human prosperity and longevity to levels never before attained, and within a period of five to six generations.

European settlement brought artisan-based industries which in turn were often replaced in the latter half of the 19th century by those with a pronounced management-labor size and separation. Cities were enlarged to provide goods and services beyond the imagination of even the first European settlers. Human comfort was enhanced by predictable internal heating systems and labor-saving kitchen appliances. Means of mobility advanced from trains and bicycles to automobiles and airplanes. Air conditioning finally created the last piece for the year-round comfort most people in the developed world take for granted. New entertainment systems and finally digital devices have made it possible for most people to have a virtual library of learning and culture at their personal disposal.

Erasing aspects of the natural world made sense within this development model. Only gradually has the wisdom of these actions been turned on its head. The natural world and its associated bio-capacity including all those annoying items listed by Dickens are now recognized for their essential role. A grudging acceptance of the balanced perspective of First Nations peoples is likewise acknowledged.

- The swamp is a wetland which not only holds water during times of active weather and flooding, but harbors places for wildlife and other living things in the natural world,

- The bush defines landscapes, provides pathways for living things to move within, and supports vegetation, allowing for carbon sequestration and climate change resilience,

- The perpetual chorus of frogs is a sign of heightened bio-diversity while their decline is an early warning of a threatened nature,

- The rank unseemly growth are those diverse natural settings free of both pesticides and a monoculture of crops allowing pollinators such as bees to flourish, and

- The unwholesome steaming earth is the mixture of fine soil and loam within which food is grown and which has significantly declined in depth since the arrival of Europeans.

Bio-capacity has multiple benefits. As well as the above-noted regulating services for climate resiliency, its bounty also includes aesthetic and cultural attributes, provisioning for food, and support for habitat enhancement, nutrient cycling, soil formation and photosynthesis. Without these services humans would have to find a means of providing them through labor and ingenuity. There is no guarantee that humanity would be successful. Today Canadians and those in advanced western societies need more bio-capacity, not less as was the case in the days of the pioneers.

The reduction in the essential ecological services of a healthy bio-capacity, has been matched by unhealthy accumulations of carbon in the atmosphere associated with emissions from buildings, transportation and various industrial processes. These correlate with rising temperatures and uncertain weather patterns. Thus today's population needs an informed environmental perspective so they have the capacity for urgent action.

Application to Education

In November 1999, a report was tabled on outreach and education in conjunction with the National Climate Change Implementation Process (Reaching out to Canadians on Climate Change, 1999). The report called for government funded outreach on the topic to several audiences including youth up to the age of 29 years and educators in order to increase support for required economic shifts. In 2003 the organization, Environmental Education Ontario (EEON) issued an appeal to address a lack of environmental education in the province of Ontario. They specifically called for strategic interventions to insure that post-secondary faculty and students were well informed about environmental issues in the province (Greening the Way Ontario Learns, 2003). In 2015 the organization issued a report card for Ontario to assess the populace's general environmental awareness and commitment to action. This report card, based on a telephone survey of 1000 Ontario residents found that 21% of the sample population were aware of issues of water pollution and 16% were aware of issues of climate change and energy. Only 6% were aware of biodiversity loss and another, perhaps the same 6%, were aware of invasive species (Omnibus Survey Report, 2015).

Since the initial 2003 report and the more recent report card, the issues of climate change, energy, and biodiversity have only become more urgent. If the world has not already gone past the 1.5 degree Celsius temperature rise goal of the Paris agreement, then it is very close to it. Once that is passed then there is only the hope of sufficient action to prevent the next threshold of two degrees. The current rise in temperature is unprecedented, not only in the short amount of time in which it has occurred but also the extent to which it correlates with human activity. Together we barely understand how it will change the world. For example, melting tundra regions would release quantities of methane with even more profound impact on climate. A severe reduction in the greenhouse gases that have caused increased temperatures is required at this moment. Likewise, we are now in what some call the 6th Great Extinction, a rapid and substantial loss of biodiversity often caused in part by loss of habitat as it is appropriated for human uses. Urgent action is required and citizens must support new policies and laws to address the potential for runaway ADC and other environmental issues.

The results of the EEON report card therefore are cause for dismay. Not enough of the population is sufficiently engaged especially if new policies appear to be costly both individually and corporately. One of the other findings of the phone survey for the 2015 report card was that people are most willing to modify their behavior towards the environment if it saves them money or contributes to their health. While those are reasonable personal motivations they appear to demonstrate a fundamental misunderstanding of the gravity of the situation. To achieve the hoped for low-carbon economy and then to engage in the process of environmental restoration the original goals of the EEON must be implemented and even furthered.

In addressing the issue of climate change and resilience, the role of educators outside of those engaged in university scientific environmental research is essential. The latter, classified as scientists concerned with either pure or applied investigation, generally appreciate the challenge ahead. The majority of educators from university lecturers, to community college faculty have the ultimate challenge of molding the critical faculties of students while embedding certain facts and ideas as givens. This can present an immediate conflict in the market driven model of contemporary academic institutions. Over the past few decades many post-secondary institutions have shifted their mandate from training up citizens who can engage critically to producing graduates who can enable economic growth. Scott (2010) believes that in order for environmental education to be successful civic values must once again be balanced against economic imperatives in order to lead students to develop cooperative behaviors and wise ethical approaches.

While the number of programs offered in Ontario with some form of environmental programming has increased over the years alongside an expansion in environmental work, students exposed to this education are primarily self-selecting, environmentally informed and motivated. It is the rest of the student population for whom there should be concern. They are led by educators who are not climate or environmental specialists, who may lack informed environmental behavior themselves, and teach in programs that ignore or may even resist a foundation of environmental knowledge. Just as an awareness of human diversity issues, and health and safety protocols, should inform all education programs, so has the environmental imperative entered this essential realm of understanding.

In a world of fake news, false hypothesis, self-interest, and corporate/government manipulation of information and conclusions, and particularly around an issue like ADC and its impact on the resilience of taken-for-granted systems from water and food to energy, how can educators overcome the biases in various points of view? How can they prepare students to reach reasonable conclusions as to the most likely explanation for the phenomenon they learn about, observe and upon which they can act? Indeed with the rise of vast amounts of information available to all on the internet the educator has the additional task of countering narrowcast media that can easily negate the content of a course attempting to lead students to an informed environmental perspective and appropriate action (Hamilton, 2011). The narrow reading of discreet facts may be a problem inviting contradictory analysis from the general scientific consensus. Thus education that enables critical thinking about all environmental knowledge is essential (Yamashita, 2015).

The word "act" is particularly significant for the manner in which it recognizes that the goal as functioning citizens is not bemoaning a world that cannot be changed. Nor can citizens blindly run off in all directions attempting to cure all maladies. Or for that matter, accept as gospel the first thing that is read or heard about an issue. But rather, to examine it as fully as possible from multiple perspectives, and to act, if such is appropriate, on this knowledge.

As well, educators need to be cognizant of the role their graduates will play. Community college educated persons, for instance, largely take up second tier positions in their respective fields. In medicine they are the nurses, first responders, and physiotherapists. In the built environment they are the technicians, technologists and operators. There is nothing pejorative about these definitions, simply an acknowledgement that ultimate decision-making resides, in the case of medicine with university trained doctors, and in the built environment with university-trained engineers and architects who have legal signing authority.

As second tier practitioners in the built environment however those technicians and technologists are effectively the life cycle guarantors of whatever sustainability features are embedded in either new or retrofitted construction. Their ongoing role contributes to both the long term financial integrity of a project but also helps realize its essential contribution to climate change mitigation and resilience.

Beyond first and second tier practitioners who work in a field normally associated with environmental issues, there are all graduates who work in other professions and fields. They may receive no education in environmental science or environmental studies unless they personally seek it in an elective. Their program may not even offer the possibility of such an elective. Thus, these students can easily go through their program and graduate without basic environmental knowledge.

The first requirement of educating young people is to prepare them for the world in which they will live. The mentoring adult has the responsibility to give them their best chance to survive and thrive in the world they encounter. Failing to do so is ethically harmful by abdicating the first responsibility of adulthood. At this point in Ontario, we are failing our students. Only a select few, those inclined towards programs that prepare them for environmental careers or recognize the multiple crises that will impact their field, are truly getting the knowledge they need for the future. In a look at communities in Brazil it was found that those who had received some education, particularly so with women, were able to better assess climate change risks and respond to them (Wamsler et al., 2012). In the end, educating people about resilience and adaptation may reduce costs for dealing with crises and disasters.

Every graduate in Ontario must be ready to live in a low carbon economy, one that must be achieved quickly. Every graduate must be prepared to deal with the crises that accompany ecosystem failure and flips, for shifts in water regimes, potential food shortages, changes in agriculture, new approaches to energy, waste management, manufacturing, living arrangements, and habitat conservation and restoration. These issues and more will have an effect on, and determine, their quality of life. They will all have to make decisions within this new context, one that might be characterized by a chaotic climate. And yet, most will graduate their college or university program this year completely unprepared. Worse, due to faculty that as of yet are not fully convinced about ADC or simply ignore the topic, these new graduates could well be hostile to the policies and legal approaches designed to engage the population in required remediation efforts.

Facing so much fundamental change in one's world can be terrifying. How then to bring along the population that will be most affected by these changes, those who will be forced to accept change and even plan for it? Post-secondary education is a last chance to really address these students. For those who will go straight to the work world and never again engage in any kind of academia, it is a final moment to grab their attention and give them the knowledge and skills they require for the future. For all those who immigrated or arrived as international students, it is a chance to inform them when they might have come from a culture without such education. For those raised in Canada and in Ontario, it is a final opportunity to shape opinions, to inform, and to encourage positive action.

Some universities and colleges have attempted to make environmental awareness more universal in their institution through their own sustainability programs. These are often programs that engage with students around issues of recycling or managing food waste. They may go further and educate around green roofs when they construct a new building, or redesign their landscape in a more naturalized manner and install bee hives. All of these initiatives are important and useful to create a context in which students are more aware of their environment and aware of useful action. However, even these initiatives do not go far enough. Stressed out students can easily walk past the information displays, ignore the call for their attention, and focus on what they perceive to be their needs. Greener universities and colleges are important but they are a start rather than an end of the work needed to prepare graduates for the world in which they will live.

There is some encouraging news about post-secondary environmental education in Canada. When students are engaged about the environment their motivations and attitudes are changed. They develop an informed environmental perspective. In 2003 a graduate student at Dalhousie University investigated the effectiveness of an Environmental Studies course offered by the Science department. She found that over the course students became more nuanced in their understanding as their knowledge deepened (McMillan et al., 2004). However, as she also notes, this group of students, like those in most environmental programs, are self-selected and generally also aware of many environmental issues. In a recent presentation members of the Canadian College and University Environmental Network (CCUEN) noted that by 2015 the Association of Universities and Colleges in Canada had listed over 200 programs focused on environmental and sustainable development in Canadian post-secondary institutions (O'Connor et al., 2015). We appear to be reaching those students who are eager to learn more, motivated to do the work, and seek to be employed in professions with an informed environmental perspective. But what about the rest of the students?

There has been much research done on learning, motivation, and how to overcome denial especially associated with climate change. This knowledge means that any curricula written for a more general post-secondary audience has to be different from the usual environmental science or even an environmental studies course. To convince students who are not drawn to environmental education of their own accord the educator has to understand the starting point of the students, their world views, and insure that the course offers opportunities for affirmation of the individual student's identity (Storksdieck et al., 2005). The course has to be something that is understood as a communication tool first, a tool to change behavior and to motivate to action. This may demand a shift in content and in the mode of presentation.

As we consider the content of a general education course in environmental issues for post-secondary education, we must first acknowledge all the good work already out there. There are programs available at many levels that address issues well. But two things suggest a re-examination of content. The first issue is the sheer urgency in what is happening to the climate and to the planet in terms of changing weather patterns, ecosystem shifts, ocean acidification, and sea level rises, species extinction, and all the effects and outcomes associated with such changes. The second issue is that environmental education must move quickly to a new audience or several new audiences, from self-selecting and already concerned to those who are in a general post-secondary program. It will require a means of addressing the whole institution through trans-disciplinary, multi-faculty cooperation and teaching (Pearson et al., 2005).

A Shift in Focus and Presentation

The challenge as citizens has become one of modifying the environmental impacts of our lifestyles with particular emphasis on reducing greenhouse gases which currently, and long into the future, will abet climate unpredictability and ocean acidification. Reducing these emissions has become an urgent public priority, alongside means for contributing to bio-capacity enhancement. Or at least this is the generally held scientific conclusion of over 95% of that community. The challenge as educators is how to communicate the problem and potential solutions to a general student population destined for a wide variety of job outcomes. Part of the task for educators is to choose how much consideration should be given to the outliers from the consensus point of view. To suggest this is a "one or the other, or a 50/50" disagreement is to unfairly load the dice in favor of those outliers whose conclusions might be one of nuance rather than opposition, or who themselves are proprietors of self-interest funding or business opportunity.

As problematic however are the recommendations for action by those who generally agree with the 95% plus scientific community. Arguments for varying types of action range on the one hand for a return to a nature more in tune with the world accepted and managed by First Nations peoples, either through willful intention, or imposed necessity, or on the other hand to an accelerated speeding up to greater prosperity and a development agenda based on the conclusion that wealthy, prosperous citizens are the best advocates for, and ingenious developers of, greener more efficient technologies and lifestyles.

If we might risk a conclusion it is that most citizens would opt for both possibilities in the range, i.e., both a heightened health and management of natural features, and increased human comfort and prosperity world-wide. The scientists might scoff at such a possibility citing entropy (all things must eventually run down and die), relativity (we've had a good run of luck and invention but even our brightest discoveries like antibiotics are short-lived respites from reality), and the deep contradictory expectations implicit in such a world view.

For a moment however let's look at its possibility from the perspective of educators rather than scientists, and particularly of students not already engaged in the conversation around environmental issues. When it comes to extending environmental education into the general population of post-secondary students, and further into the adult population, two issues arise. First is the issue of content. What content is useful to educate, inform, encourage action and motivate the desired behavior? The second issue is presentation. How should such a curriculum be structured and presented? Is it a kind of dumbed down shadow of most sustainability or environmental science courses? Or should it be something different reflecting the nature of the audience? This audience is critical. The various graduates of the many diverse programs across the province and Canada will engage in important decision making during their lifetime in response to environmental issues. Their shopping choices will influence how food is produced, how water is managed and what kind of energy is used to produce it. Their decisions and commitment to waste management will influence agriculture, processing and manufacturing. What they learn and how they learn is important. Arguably, it is this yet to be fully engaged audience who will have the most impact on how Canadians adjust systems and infrastructure to respond to ADC and other environmental issues.

Environmental education has been around for some time, long enough for research to have been done on what works best. Accordingly the UNESCO report of 2012 that concluded the Decade of Education for Sustainable Development comments that ultimately a variety of teaching approaches appears to elicit desired behavior. Responses to their research demonstrated that "all seem to point to a need for well-rounded, interactive, integrated and blended forms of learning" (Shaping the Education of Tomorrow, 2012). It is imperative to begin by understanding the audience, their beliefs, attitudes and values and how people construct their attitudes about climate change and other environmental issues (Brownlee et al., 2013). Pedagogical methods that stimulate self-determined behavior also appear more successful in motivating students to make better decisions regarding the environment (Darner, 2009). Newman and Fernandes (2016) tested and found a wide range of variables within different contexts influence behavior towards environmental issues which continues to point towards educational methods that account for multiple influences on the student's decision making.

However, while education overall correlates with motivated and informed care for the environment in students and adults (Meyer, 2015), increased knowledge is not necessarily the answer for the general population. It is an easy assumption by educators that it is simply lack of information that causes poor responses. Accordingly, most educators seek to inform first. But much research demonstrates that people's decisions are based on a host of variables that have little or nothing to do with knowledge acquisition and that those variables tend to be weighted differently when it comes to decision making (Heeren et al., 2016; West, 2015). Beyond that, there is a reason these students are not in environmental programs. They may not be interested but sometimes it is due to an aversion to the required material especially science and mathematics. And these may be key differences between the self-selected population who sign up for environmental programs and courses and those who do not. Attempting to meet this larger and crucial general population of students requires a re-evaluation of pedagogical approach and materials.

As the government moves forward in making the required changes to promote a low carbon economy it needs the general population on board with the policy shifts. However, research shows, that with adults their world view often outweighs knowledge when it comes to support for such policies. And that additional knowledge can actually increase polarization and resistance. Yet, young people, teens and perhaps young adults, are still formulating their world view and are more open to change (Hobson and Niemeyer, 2012; Stevenson et al., 2014). This means that addressing the post-secondary population not engaged in environmental programs may be critical for future policy support.

Research has gone into determining how to motivate behavior in caring for the environment. But aside from campus wide initiatives that often involve food waste or energy use, little has been asked about the general population of students not involved in environmental programs and career choices. However, research into the factors that influence decision-making demonstrate that people are led by everything from their own sense of self-determinism and self-affirmation (Sparks et al., 2010), to peer and social pressure (Gifford and Nilsson, 2014), to a context of collective action or its lack (Pongiglione, 2014; Sweetman and Whitmarsh, 2016; Obradovich and Guenther, 2016), or as is often described as intrinsic and extrinsic motivations to moral emotions in general (Xie et al., 2015), to the morality inherited from the parenting style of their family (Barker and Tinnock, 2006) and finally, to a need to justify existing systems (Feygina et al., 2010). The general population of students on a Canadian campus today brings all of these variables to a course on the environment. They come from varied backgrounds, diverse countries and cultures, different religions and ethnicities. All of these influence how they perceive their own and collective responsibilities towards the environmental context in which they live. Course content and teaching approaches must account for these variables to be effective (Heimlich and Ardoin, 2008; Brownlee et al., 2013; Klockner, 2013).

Course content may be able to address some of these variables. Certainly, environmental science and studies classes include components on morality and environmental justice for human beings and other species. But this material would have to be tailored to a population where the educator cannot assume a starting point of general agreement. Thus the course content needs to provide a means, perhaps subtly, of speaking to the often hidden factors that impede both learning and sustained post-course behavior. It is for that reason that it may be useful to shift the focus in a general education course from sustainability to resilience and regeneration. These concepts, resilience and regeneration, may better address those students with more conservative moral frameworks. Resilience speaks of strength, both individual and community and the ability to withstand hardship and difficulties. Regeneration addresses the nostalgia often a part of a conservative moral framework. It seeks to restore what has been lost. This content shift is also much more in tune with the current situation, the urgency of swiftly shifting ecosystems and water regimes along with the mandate for both mitigation and adaptation.

One way to measure resiliency is to compare bio-capacity to carbon footprint. First there's some apparent good news at least from a Canadian perspective. Unlike virtually every country with an advanced industrial economy Canada has the distinction of having twice as much bio-capacity in comparison to its ecological and carbon impact. Why? Simple answer—it's a big country with a relatively small population. On a per person basis however each Canadian is as profligate, or more so, than the highest consuming polluters in the world.

So is there a Canadian model that might provide a way to tackle the current challenge of advancing prosperity while at the same time lowering environmental impact? Canadian bio-capacity as measured against the absolute Canadian ecological/carbon footprint is at a ratio of roughly 2:1, providing perhaps an inadvertent but useful heuristic as a go-forward approach.

We need to be careful however about suggesting that this is a scientific conclusion. The 2:1 is a loose and crude calculation that comes from a particular set of data points put together by the Global Footprint Network. They use over 200 data points to calculate a national ecofootprint and the biocapacity to sustain the population under the current management practices. They have data that starts in the 1960's.

The problem for Canada is that its calculation ratio has been decreasing since then. Canada has not held steady like Finland (at around 2.27:1). Instead the 2:1 is actually a less than stellar outcome born of poor management of resources. In other words, if the world is looking for good models Finland is much better than Canada.

The Canadian ratio is based on 2012 data so it has likely got worse since then. It does retain currency however as a target to maintain and in some measure to regain, or as a model to be approached for greater certainty on a scientific basis. Perhaps more useful however is relying on this simple heuristic as a marketing tool—a way to organize and mobilize a population. It provides a simple, easily understandable goal to mobilize the general population of students and graduates. This framing in terms of motivation appears important. A study in Ontario demonstrated that, more so for men than for women, framing the issue of climate change as a motivational message rather than one of sacrifice elicits a more positive response (Gifford and Comeau, 2011). Negatives do not motivate people well. The real use of the 2:1 ratio is perhaps as a motivator for changes in human behavior, beginning in Canada but also as a goal in other developed countries.

This is the problem with using footprint calculators in academic institutions. We are essentially handing students a negative number. Even one of the authors of this article, a resident of a solar powered 'green' house, with a commitment to transit use, and reduced meat consumption still comes out at three planets worth of carbon footprint. A significant aspect of this is simply how hard it is to reduce carbon footprint within the social and physical structures of where most people live in the developed world.

One of the ways to motivate climate change deniers and the politically adverse, those that tend toward a conservative morality, may be to focus on how much has been lost and the need to restore what was once plentiful or healthy in the past, that is, a conservation approach. This is actually what the ratio measures. It shows what we had in the past and what we are losing. Restoration is what we are looking for first. Research has shown that while the moral framework for those who lean politically conservative includes in-group identity, compliance and purity, such people are really not that different from those who are more progressive politically. Both groups seek to reduce and mitigate harm (Schein and Gray, 2015). Content in general education courses should be shifted to reflect this desire in all students to assess and then reduce harm to themselves, their loved ones, communities and society as a whole. Pushing further in terms of ethics, the ratio could then be adjusted for an optimum regeneration level and restoration of biodiversity, though more work needs to be done on what that would be and from where the data would come. Concurrently, erroneous approaches and false facts that produce skepticism should not be affirmed as it can cause confusion in the classroom. Critical thinking and constant references to scientific consensus are recommended by Torcello to counter such skepticism (2016).

There is one large gaping hole in the current data sets measured by the Global Footprint Network. Their measure of biocapacity is purely human centered. They measure the capacity of local fishing grounds, inland fish production, forest production, and cropland to satisfy the needs of the population. There is no measure of biodiversity or any kind of measure for the satisfaction of the needs of other species. In terms of regeneration, this is a significant problem. In other words, using their data, the heuristic of 2:1, may fall short of being truly regenerative.

Recognizing the tentativeness of the above conclusions, it is suggested that environmental education be shifted to emphasize resilience and regeneration. Students could be taught that for each unit of ecological/carbon footprint humans expand by their activities they should as a beginning point and at minimum add two units of bio-capacity. Means range from restoring, upgrading, renewing or creating the conditions, places, and investments for this action. A simple action to increase bio-capacity is to plant a tree, this helps mitigate greenhouse gas emissions, deals in urban areas with the heat island effect, manages rainfall, and provides habitat for birds and perhaps other species. A naturalized yard is similar in its outcomes. Collective actions can deal with areas and issues that are larger in scope. Yet the motivation is rewarded with a positive outcome rather than the struggle of trying to reduce a carbon footprint within a cultural context which can make that very difficult to accomplish. Additionally, such actions are real and measurable outcomes to environmental education. As Heimlich advocates, rather than measure individual knowledge and attitudes, the real shift is in the measurable conservation actions that make a difference (2010).

This approach acknowledges two realities. On the one hand the world's ecological/carbon footprint will increase for the simple reason that despite reduced per-unit energy use in the way modern cars operate and building efficiencies are achieved, cumulative intensification is inevitable in a world of increasing population and an improving quality of life with associated middle class consumer expectations. As well, the mixture of sources of carbon release and the environmental impact of continuing resource use will likely retain a similar character as today into the foreseeable future. This will occur even with the use of alternate renewable energy means. Wind turbines and solar collectors after all require the use of rare earth and other minerals and materials, a construction process, transportation to a site, mounting, maintenance and finally the eventual disposal, all with degrees of carbon and ecological impact.

Countering this is the necessity therefore not simply for bio-capacity replacement through a series of one-for-one measures but its increase as a function of enhancing and increasing natural resources fecundity and associated ecological services. This creates ecosystem redundancy, a necessary component of resilience and regeneration. This takes students beyond discussions of sustainability and onto action that can both mitigate and enhance adaptation to climate change.

Such an approach actually takes seriously the current situation. It surpasses the checklist approach of so much sustainability programming which is based on reducing impact, or simply slowing the rate of decline often through perceived or real sacrifice. It is a net positive development philosophy that responds to a real potential catastrophe rather than one playing about the edges for the feel good sense of "well at least we tried!"

Tools and protocols are required which link bio-capacity increase alongside inevitable carbon/ecological impact and ones which refine the ability to measure these initially on a 2:1 basis. A necessary step is appreciating the accordion-like nature of this examination, i.e., as more eco/carbon impact occurs the accordion is necessarily much wider necessitating a doubling in the size of the bio-capacity piece, but as efficiency and less impactful means of meeting human comfort are implemented, and which possibly conserve or add to bio-capacity, a narrower accordion is required in which the bio-capacity enhancement though smaller is more strategically aimed at crucial points of repair or replenishment. Students will need to understand that resilience 2:1 is a process dependent upon its context.

Ironically this approach may be the best resilience strategy for coping with climate unpredictability as well. The re-institution of aspects of an enhanced bio-capacity provides measures for protection from violent weather conditions, cleans the air, provides shade, stores water, contributes to food provisioning, and acts to mitigate climate change causal factors by sequestering carbon. Informed environmental behavior based on resilience and regeneration offers the best chance for an individual or community to make an essential difference.

For decades sustainability and sustainable development have been the foundations of many of the university and college environmentally focused programs. This creates potential problems when approaching a general education course and in preparing young adults for the future they face. Resilience is based on disaster preparation. It acknowledges the crisis facing humans and other species and seeks not just mitigation but adaptation. Such content may actually contradict what has been taught in sustainability courses. At this point, in light of reality, environmental education courses as they have been taught may actually impede required preparation in resilience (Krasny and DuBois, 2016; Lundholm and Plummer, 2010). This shift in content is one that honestly acknowledges the new reality and has the possibility to address at least some of the variables that motivate human behavior. It has the possibility of engaging the general population of students who have shown no interest in an environmental professional career, or are frustrated with the sense of doom, have little information on how to positively respond, and possibly are science adverse.

Conclusion

In conclusion, what does this mean for the educator?

One of the main issues regarding climate change preparedness and associated resilience planning is the lack of understanding of basic climate change issues in the population. Most data is too complex to motivate changes in behavior. And there is a crucial population sitting in classes every day who desperately need to understand the situation and have a positive motivator given to them to insure changes in daily life and the required social, economic and political action.

Beyond that, while some planners and cities or subnational entities have the capacity to deal with complex data production and application, many do not. The heuristic ratio of 2:1 could be developed so that smaller groups right down to the neighborhood level have a means of analysis and action within their own locales. 2:1 is a tool that needs to be developed for particular audiences to achieve particular results and, above all, become a foundation for action to mitigate climate change as well as to create greater resilience at all scales.

The heuristic 2:1 should be developed as a motivational tool to promote action. However, with one significant change, that is, to add in some sort of biodiversity indicators beyond those simply in service to humankind. Through the heuristic of 2:1, information can be created and formatted for use in college and university classrooms as well as for the use of planners, and smaller scale polities.

The latest innovations in communication through mobile apps and social media need to be employed to communicate this tool as means for popular education and motivation regarding climate change. This is the crucial need right now. Canada needs an educated public in order to support the changes required through policies and legislation not only to return to a 2:1 ratio but then to better it. 2:1 is a starting point not the end goal.

References

Barker, D. and Tinnick, J. III. 2006. Competing visions of parental roles and ideological constraint. *American Political Science Review* 100(2): 249–263.

Brownlee, M., Powell, R.B. and Hallo, J.C. 2013. A review of the foundational processes that influence beliefs in climate change: opportunities for environmental education research. *Environmental Education Research* 19(1): 1–20.

Dickens, C. 1842. American Notes for General Circulation, Chapter 13, Chapman and Hall, London U.K.

Darner, R. 2009. Self-determination theory as a guide to fostering environmental motivation. *The Journal of Environmental Education* 40(2): 39–49.

Feygina, I., Jost, J.T. and Goldsmith, R.E. 2010. System justification, the denial of global warming, and the possibility of "system-sanctioned change". *Personality and Social Psychology Bulletin* 36(3): 326–338.

Gifford, R. and Comeau, L. 2011. Message framing influences perceived climate change competence, engagement and behavioral intentions. *Global Environmental Change* 21: 1301–1307.

Gifford, R. and Nilsson, A. 2014. Personal and social factors that influence pro-environmental concern and behavior: A review. *International Journal of Psychology* 49(3): 141–157.

Global Footprint Network, accessed 8 January 2016. http://www.footprintnetwork.org/en/index. php/GFN/.

Greening the Way Ontario Learns. A Public Strategic Plan for Environmental and Sustainability Education, 2003. EEON, Ontario, Canada.

Hamilton, L. 2011. Education, politics and opinions about climate change evidence for interaction effects. *Climate Change* 104: 231–242.

Heeren, A.J., Singh, A.S., Zwickle, A., Koontz, T.M., Slagle, K.M. and McCreery, A.C. 2016. Is sustainability knowledge half the battle? An examination of sustainability knowledge, attitudes, norms, and efficacy to understand sustainable behaviors. *International Journal of Sustainability in HIgher Education* 17(5): 613–632.

Heimlich, J. and Ardoin, N.M. 2008. Understanding behavior to understand behavior change: a literature review. *Environmental Education Research* 14(3): 215–237.

Heimlich, J. 2010. Environmental education evaluation: Reinterpreting education as a strategy for meeting mission. *Evaluation and Program Planning* 33(2): 180–185.

Hobson, K. and Niemeyer, S. 2012. What skeptics believe: The effects of information and deliberation on climate change skepticism. *Public Understanding of Science* 22(4): 396–412.

Klockner, C. 2013. How powerful are moral motivations in environmental protection? an integrated model framework. pp. 447–472. *In*: Heinrichs, K., Oser, F. and Lovat, T. (eds.). Handbook of Moral Motivation: Theories, Models, Applications. Sense Publishers, Rotterdam, Netherlands.

Krasny, M. and DuBois, B. 2016. Climate adaptation education: embracing reality or abandoning environmental values. *Environmental Education Research*. DOI: 10.1080/13504622.2016.1196345.

Lundholm, C. and Plummer, R. 2010. Resilience and learning: a conspectus for environmental education. *Environmental Education Research* 16(5-6): 475–491.

McMillan, E., Wright, T. and Beazley, K. 2004. Impact of a university-level environmental studies class on students' values. *The Journal of Environmental Education* 35(3): 19–27.

Meyer, A. 2015. Does education increase pro-environmental behavior? Evidence from Europe. *Ecological Economics* 116: 108–121.

Muzyka, K. 13 January 2017. Trudeau's 'phase out' oil sands comments spark outrage in Alberta. If [Trudeau] wants to shut down Alberta's oilsands ... he'll have to go through me, says Wildrose leader. CBC News, http://www.cbc.ca/news/canada/edmonton/justin-trudeau-oilsands-phase-out-1.3934701.

Newman, T. and Fernandes, R. 2016. A re-assessment of factors associated with environmental concern and behavior using the 2010 general social survey. *Environmental Education Research* 22(2): 153–175.

Obradovich, N. and Guenther, S.M. 2016. Collective responsibility amplifies mitigation behaviors. *Climate Change* 137: 307–319.

O'Connor, T., Westman, M. and Wright, T. 2015. Environment and Sustainability Studies in Post-Secondary Education: Reflections on the Past, Present and Future. CCUEN, Canada.

Omnibus Survey Report. 2015. EEON, Ontario, Canada.

Pearson, S., Honeywood, S. and O'Toole, M. 2005. Not yet learning for sustainability: the challenge of environmental education in a university. *International Research in Geographical and Environmental Education* 14(3): 173–186.

Pongiglione, F. 2014. Motivation for adopting pro-environmental behaviors: the role of social context. *Ethics, Policy & Environment* 17(3): 308–323.

Reaching Out to Canadians on Climate Change, A Public Education and Outreach Strategy. 1999. Submitted as part of the National Climate Change Implementation Process, Canada.

Schein, C. and Gray, K. 2015. The unifying moral dyad: liberals and conservatives share the same harm-based moral template. *Personality and Social Psychology Bulletin* 41(8): 1147–1163.

Scott, D. 2010. Preparing future generations: climate change, sustainability and the moral obligations of higher education. *The Ometeca Journal* XIV-XV: 53–76.

Shaping the Education of Tomorrow. 2012. Report on the UN Decade of Education for Sustainable Development, Abridged. UNESCO, Paris, France.

Sparks, P., Jessop, D.C., Chapman, J. and Holmes, K. 2010. Pro-environmental actions, climate change, and defensiveness: Do self-affirmations make a difference to people's motives and beliefs about making a difference? *British Journal of Social Psychology* 49: 553–568.

Stevenson, K.T., Nils Peterson, M., Bondell, H.D., Moore, S.E. and Carrier, S.J. 2014. Overcoming skepticism with education: interacting influences of worldview and climate change knowledge on perceived climate change risk among adolescents. *Climate Change* 126: 293–304.

Storksdieck, M., Ellenbogen, K. and Heimlich, J.E. 2005. Changing minds? Reassessing outcomes in free-choice environmental education. *Environmental Education Research* 11(3): 353–369.

Sweetman, J. and Whitmarsh, L.E. 2016. Climate justice: high-status ingroup social models increase pro-environmental action through making actions seem more moral. *Top. Cogn. Sci.* 8: 196–221.

Torcello, L. 2016. The ethics of belief, cognition, and climate change. pseudoskepticism: implications for public discourse. *Topics in Cognitive Science* 8: 19–48.

Wamsler, C., Brink, E. and Rentala, O. 2012. Climate change, adaptation and formal education: the role of schooling for increasing societies' adaptive capacities in El Salvador and Brazil. *Ecology and Society* 17(2): 2. http://dx.doi.org/10.5751/ES-04645-170202.

West, S. 2015. Understanding participant and practitioner outcomes of environmental education. *Environ. Educ. Res.* 21(1): 45–60.

Xie, C., Bagozzi and Grønhaug, K. 2015. The role of moral emotions and individual differences in consumer responses to corporate green and non-green actions. *Journal of the Academy of Marketing Science* 43: 333–356.

Yamashita, H. 2015. The problems with a 'fact-focused' approach in environmental communication: the case of environmental risk information about tidal flat developments in Japan. *Environmental Education Research* 21(4): 586–611.

Index

Color Plate Section

Chapter 4

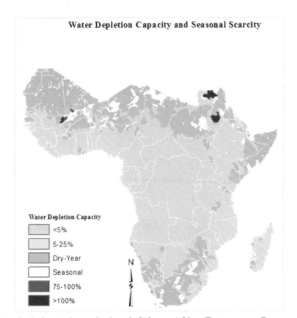

Figure 4.1 Water depletion and scarcity in sub-Saharan Africa (Data source: Brauman et al. (2016)).

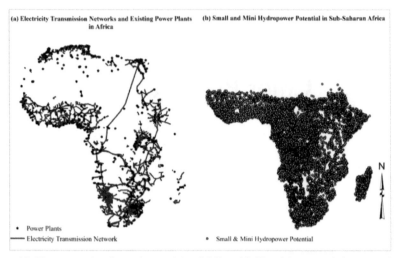

Figure 4.2 The energy situation and potentials of Africa: (a) Electricity transmission networks and existing power plants in Africa; and (b) Potential for small and **mini-hydro power** production in Sub-Saharan Africa (Data sources: the Africa Electricity Transmission Network (AICD 2009) and studies carried out by Dimitris Mentis of the Division of Energy Systems Analysis, Royal Institute of Technology in Stockholm, Sweden).

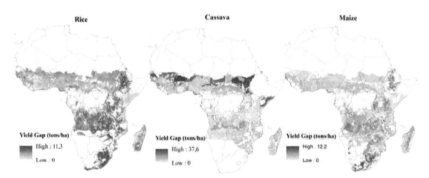

Figure 4.3 Yield gaps for major food crops in Sub-Saharan Africa based on growing degree days and precipitation (Data source: Foley et al. (2011) and Mueller et al. (2012)).

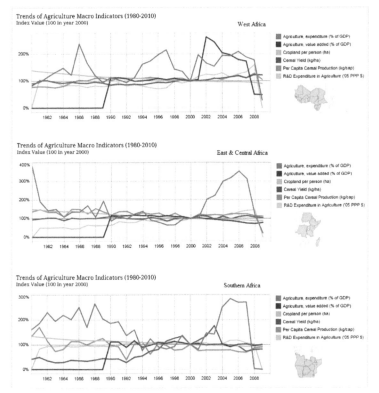

Figure 4.4 Trends of agricultural macro indicators for different regions of Africa (1980–2010) (Data source: FAOStat, ReSAKSS, ASTI, IFPRI, HarvestChoice).

Chapter 6

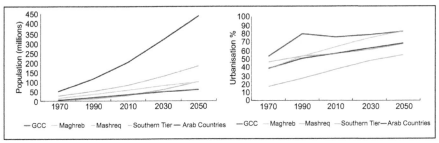

Figure 6.2 A diagram of the Arab region's urban population and urbanization trends (1970–2050) (UN-HABITAT, 2012; UNDESA, 2011).

Chapter 12

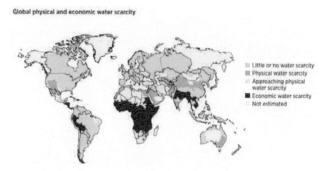

Global physical and economic water scarcity

Little or no water scarcity
Physical water scarcity
Approaching physical water scarcity
Economic water scarcity
Not estimated

Figure 12.1 World water scarcity.[xvi]

Chapter 13

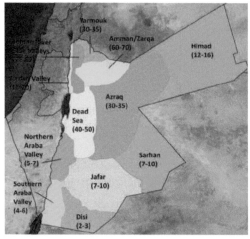

Figure 13.1 Groundwater Basins in Jordan and their annual safe yield in million cubic meters.

Printed and World Water Development Report 30 March 2014. World Water Assessment Programme (WWAP), As cited in: UN water (2014/11/24) Water Stress versus Water Scarcity, Available at: http://www.un.org/waterforlifedecade/scarcity.shtml (Accessed: 08.04.2016).

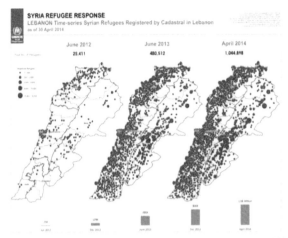

Figure 13.2 Distribution of Syrian refugees in Lebanon (UNHCR, 2014).